Thrive in Bioscience | Revision Guides

Thrive in Biochemistry and Molecular Biology

Lynne S. Cox
Department of Biochemistry, University of Oxford and
George Moody Fellow in Biochemistry, Oriel College, Oxford

David A. Harris
Department of Biochemistry, University of Oxford and
E. P. Abraham Cephalosporin Fellow in Biochemistry, St Anne's College, Oxford

Catherine J. Pears
Department of Biochemistry, University of Oxford and
Fellow of University College, Oxford

Thrive in Bioscience | Revision Guides

UNIVERSITY PRESS

OXFORD
UNIVERSITY PRESS

Great Clarendon Street, Oxford, OX2 6DP,
United Kingdom

Oxford University Press is a department of the University of Oxford.
It furthers the University's objective of excellence in research, scholarship,
and education by publishing worldwide. Oxford is a registered trade mark of
Oxford University Press in the UK and in certain other countries

© Lynne S. Cox, David A. Harris, and Catherine J. Pears 2012

The moral rights of the authors have been asserted

Impression: 1

British Library Cataloguing in Publication Data
Data available

Library of Congress Cataloguing in Publication Data

Library of Congress Control Number: 2012933760

ISBN 978-0-19-964548-0

Printed in Great Britain by
Ashford Colour Press Ltd, Gosport, Hampshire

Contents

Contents

Chapter 5

Chapter 6

Chapter 7

Chapter 8

Chapter 9

Four steps to exam success

1 Review the facts

This book is designed to help your learning be quick and effective:

- Information is set out in bullet points, making it easy to digest
- Clear, uncluttered illustrations illuminate what is said in the text
- Key concept panels summarize the essential learning points

2 Check your understanding

- Try the questions at the end of chapters and online multiple-choice questions to reinforce your learning
- Download the flashcard glossary to master the essential terms and phrases

3 Take note of extra advice

- Look out for revision tips, and hints for getting those precious extra marks in exams

4 Go the extra mile

- Explore any suggestions for further reading to take your understanding one step further

Go to the Online Resource Centre for more resources to support your learning, including:

- Online quizzes, with feedback
- A Flashcard glossary, to help you master the essential terminology

online resource centre

www.oxfordtextbooks.co.uk/orc/thrive/

Using this guide

This book is designed to be an *aide memoire* and a quick check-up source rather than an authoritative text. You will need to have revised material from your lecture notes, textbooks, and recommended reading lists in order to get full value from this book. We have tried to include core material that is central to an understanding of biochemical processes. However, courses differ, so not all the material here may be covered in your course, and there may be aspects we haven't included here that you do need to know about. It is also worth checking out the other revision guides in this series on, for example, cell biology and genetics, according to your own course requirements.

The material is arranged into related sections, though of course there is cross-talk between aspects of biochemistry in the cell, so we have tried to cross-reference sections where relevant. Note that we have used italics to highlight enzymes—this does not imply that enzyme names are normally italicized, but it should help you to be able to identify the relevant enzyme in the general body of the text. We have highlighted important terms in bold and included a definition of these in the Glossary.

Sometimes it can be hard to remember all the points about a topic. We have broken the topics down into smaller chunks to make revision easier. In addition, it can help to learn the number of steps in a biochemical pathway, e.g. if you know that there are ten steps in glycolysis, this should help you to ensure that you don't miss out anything crucial.

Exam technique

When answering **short questions**, ensure that your answer is concise and precise, and that it contains all the information required but no waffle or irrelevant information. In some cases, it is OK to use bullet points, but not in others, so make sure you choose a format appropriate to your own university's exams. Although exam styles vary, examiners will always appreciate scientific accuracy, relevance, and logical argument.

We have included some sample exam questions in the book, with hints (we haven't given you worked answers as the text itself is in essence the answer—we have boiled the information down to the bare minimum). There are also multiple choice tests online linked to each section of the book, so you can check your progress in revision.

 online resource centre Go to www.oxfordtextbooks.co.uk/orc/thrive/

Using this guide

For **multiple choice questions**, try to come up with the correct answer first, then check the answers given—be particularly careful when you have to identify correct combinations of answers.

---

e.g. Which of the following is true about eukaryotic translation?
   i)   The ribosome is made up of 40S and 60S subunits forming a 100S complex.
   ii) The ribosome is made up of 40S and 60S subunits forming an 80S complex.
   iii) The AUG start codon of the mRNA is initially located in the P site of the ribosome.
   iv) The AUG start codon of the mRNA is initially located in the A site of the ribosome.
   v)   Peptide bond formation is mediated by an RNA enzyme (a ribozyme).

Answers:

A i, iv, v   B i, iii, v   C ii, iii, v   D ii, iv, v   E ii, iv

(the correct answer is C)

When tackling **exam essays**, remember:

- Read the question carefully, and try to identify exactly what the examiner is asking for (e.g. is the question limited to prokaryotes or eukaryotes, or should you include both?). Are details of a single biochemical process required, or do you need to integrate material from various parts of your course to answer the question fully? If so, do you know all those topics well enough to provide a balanced answer?

- For a one-hour essay, spend at least five minutes planning at the start—it may seem scary when everyone else appears to have launched straight in, but your essay will be coherent and well-structured, and planning makes sure that you don't forget to include critical points, and that you can balance the material appropriately. Preparation is key here—it really helps if you have already thought out various different essay plans for every conceivable way a question could be asked on each topic, as this saves a lot of time in exams; the pressure is much less if you are confident that you have prepared thoroughly. There is no single correct way to plan an essay—some people make mind maps, others prefer lists, but do make sure that you don't waste time writing full prose (save that for the essay; you can use abbreviations as much as you like in a plan). As you are writing, do keep checking your plan to make sure you include everything you think important.

- Avoid waffle—every word needs to count so make sure it's scientific and conveys information quickly, clearly, and coherently. Do make sure you use technical terms correctly and spell scientific words carefully. Examiners are not just looking for factual content but evidence that you understand the material. You can demonstrate this by using a clear essay structure (subheadings can really help guide the examiner in your thought·processes—check if it's permitted at your institution) and by providing a cogent argument.

- Try to include an introduction, the core material, and if time, a rounded conclusion …
- The introduction should contain a definition of the key terminology of the question at the start of your answer. You can pick up marks quickly this way and begin the essay in a biochemically convincing manner. For example, 'DNA replication is the template-directed synthesis of a polymer of deoxyribonucleotides mediated by DNA polymerases involving formation of phosphodiester bonds, and is driven by a highly negative ΔG from PP_i hydrolysis …' sounds rather more scientific than 'DNA replication is important and happens in all cells every time they divide'—a statement that is equally true but much less biochemically precise. You can also set the scope of the essay, e.g. 'I shall discuss predominantly eukaryotic ATP synthesis pathways, but will touch on those in prokaryotes where appropriate …'

Paragraphs in the **main body** of the essay should contain:

- Key **concept**, with details of biochemical process (as relevant).
- Specific **named example(s)**.
- **Experimental evidence** to support the ideas (even when not specifically asked for in the question). By providing support, you show the examiner that you understand how information was obtained and how strong the evidence is behind the idea—it's a great way of bumping up marks. As you progress through your degree, you will be expected to refer directly to experiments in the primary research literature, so get into the habit of providing experimental evidence early on and it will make it much easier later.
- If possible, illustrate with an **annotated diagram**, e.g. draw out the biochemical pathway (SIMPLY but clearly—you are being marked on scientific content not artistic merit). Remember that diagrams are only of exam value if they are fully annotated (i.e. have descriptive labels) and **don't take longer to draw than it would have taken you to write text describing them**.
- In most universities, it is fine to use colour in diagrams, but do check first. For example, illustrating the complex processes of DNA recombination is made much simpler by using a different colour for each strand of DNA. We have tried to make the diagrams in this book simple and easy to reproduce under exam conditions (though we couldn't use colour because of printing constraints). Most undergraduate textbooks employ graphic artists to make the pictures beautiful; you won't have time to do this in an exam! For example, it is more informative to draw the relevant components of the lac operon as boxes and use good annotation to explain what is happening under different sugar conditions, rather than spending time making your DNA into beautiful double

helices but missing the scientific point. (Obviously if the essay asks about the structure of DNA, then you have to draw the double helix not boxes!).

- Don't repeat the same material in diagrams and text, but do introduce each diagram, e.g. 'the subunit composition of the eukaryotic ribosome is shown'. Remember to use full labels that are clear, legible, and informative.

- Ensure that you leave sufficient space around your diagrams so the reader can easily see what you are illustrating—don't squeeze them into the margins or wrap text around them.

- Comparative tables can also be really helpful to convey factual information rapidly and show the examiner that you can identify key concepts and relate the details to those concepts (e.g. concept: initiation, elongation, and termination in DNA replication; details: the analogous prokaryotic and eukaryotic proteins involved in each step), but make sure you don't rely wholly on such tables, as you will also be marked on the coherency of your discussion.

- Round off each paragraph with how the information you have just presented addresses the question.

Try to include a **conclusion** where you can argue for/against (especially if it's a 'Discuss ...' type essay). You can throw in the odd quirky example here if it didn't fit well into the rest of the essay, but make sure it's relevant. If there are controversies in the field, you can mention them here. Don't waste time repeating things you have already mentioned. You could even highlight what further knowledge is required to fully understand the process (but make sure it's a real gap in knowledge, not simply that you didn't know it!).

Good luck in your exams and do remember that biochemistry is a subject to enjoy—it's not just about passing tests!

Lynne S. Cox
David A. Harris
Catherine J. Pears

Oxford
March 2012

Acknowledgements

All the authors wish to thank their long-suffering families, putting up with the stresses and strains that are intrinsic in writing a book, and, in particular, Sophie for her extensive help in sorting out the glossary. So heartfelt thanks to Charlie, Sophie and Lucy; Carol, Ben, Dan and Sam; Nigel, Hilary, Eliza and Sam for their patience.

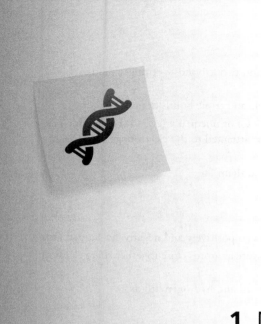

1 Molecules

1.1 BONDS

Living organisms are made of up of organic molecules consisting mainly, though not exclusively, of carbon (C), hydrogen (H), oxygen (O), nitrogen (N), and sulphur (S). These are held together by bonds which vary in strength and length.

Covalent bonds

- **Covalent bonds** form when electrons are shared between atoms within a molecule
- These are usually the most stable type of bond, although some will break spontaneously
- e.g. C–C or C–H
- Can be single, double, or triple bonds, e.g.

H_3C —— CH_3 H_2C === CH_2 HC ≡≡≡ CH

   ethane   ethene   ethyne
                     (ethylene)   (acetylene)

    i. sigma (σ): single bond, strongest, formed by head-on orbital overlap, symmetrical with respect to rotation around bond axis
    ii. pi (π): double bond where two lobes of one atomic orbital of one atom overlap two lobes of the other, weaker than σ, on-rotational

1

iii. triple bonds formed by one σ- and two π-bonds—stronger than each individual bond
- More energy is required to break double and triple bonds than single bonds
- Electrons can be shared equally (e.g. C–C) or unequally (e.g. O–H) in which case the bond is polarized with the electrons attracted to the electronegative O atom (forming a dipole)
- Many require **enzymes** to make or break them

Ionic bonds

- **Ionic or electrostatic bonds** form between positively and negatively charged ions
- Have some degree of covalent bond nature as atoms close together share electron density
- Intermediate in strength between covalent and hydrogen bonds
- e.g. Na^+Cl^-

Dipole–dipole interactions

- Dipole = separation of positive and negative charge
- Interaction between permanent dipoles increases **attraction** between molecules
- e.g. HCl

$$\delta+ \quad \delta- \qquad \delta+ \quad \delta-$$
$$H\!-\!\!-\!\!Cl\cdots\cdots H\!-\!\!-\!\!Cl$$

Hydrogen bonds (H-bonds)

- **Hydrogen** or **H-bonds** form between polar molecules
- When H is attached to an electronegative atom such as O or N, the bond becomes polarized
- Bond formed by sharing of non-bonding (lone pair) electrons from one atom with an H covalently attached to an electronegative atom and therefore starved of electrons
- Weaker than ionic bonds but stronger than hydrophobic or van der Waals forces—essentially a strong dipole–dipole interaction
- e.g. stabilize secondary structures of proteins (alpha helices and beta sheets), base pairs in DNA double helix and between water molecules (as shown below)

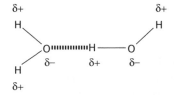

van der Waals forces

- **van der Waals forces** occur between any atoms
- Transient dipoles formed by electron movement lead to electrostatic **attraction** between atoms/molecules a short distance apart
- Can form between polar or hydrophobic molecules
- Individually weak but additive in large molecules
- At very short distances, van der Waals interactions become strongly repulsive (≤atomic radius)

Hydrophobic effect

- The **hydrophobic effect** will drive the association of hydrophobic molecules in a polar, aqueous environment to exclude water and maximize entropy
- e.g. fatty acids in centre of lipid bilayer or fat droplet
- Individually weak but additive in large molecules

Comparison of bond types

For comparison of properties of important bond types see Table 1.1.

Bond	Approx. strength (kJ/mol at 25°C)	Length (internuclear separation distance) (Å)	Where found	Example
Covalent	200–1000	0.7–2.7, e.g. 1.4 for C–C in graphite	intermolecular	peptide bond
Ionic or electrostatic	~40	2.8	inter- and intramolecular	salt bridges
Dipole–dipole	4–20		inter- and intramolecular	
Hydrogen	10–30	2.7–3.1	inter- and intramolecular	between polar side chains of amino acids
van der Waals	4	3–4	inter- and intramolecular	between tightly packed atoms in centre of protein molecule
Hydrophobic effect	<1 per CH_3	3–5	inter- and intramolecular	fatty acids in lipid droplet or membrane

Table 1.1 Comparison of bond types
Note: strength is approximate and varies according to molecules involved.

 Check your understanding

Describe the different bond types occuring in organic molecules. (*Hint: can you give an example of each type?*)

1.2 PROTEINS

Proteins are essential to all living organisms as they catalyse the majority of enzymatic reactions in the cell and also play an essential structural role. Proteins are polymers of amino acids also referred to as polypeptide chains. The function of a protein is intimately linked to its structure.

Key functions of proteins

- Structural, e.g.
    i. collagen, which forms long fibres
    ii. actin, which forms long filaments made up of short monomers
- Enzymes, e.g. lysozyme
- Carriers, e.g. haemoglobin
- Transmembrane, e.g. ion channels
- Signalling, e.g. insulin

Amino acids

- Fundamental building block of **proteins**
- General structure of an **amino acid** (shown in non-ionized form below and as **zwitterion**)

- Amino acids are **chiral** about the central carbon atom next to the COOH (α-carbon)
- Natural amino acids are the L form (laevorotatory)
- Twenty naturally occurring in proteins that differ in the R side chain (Table 1.2)
- Side chains confer one or more specific characteristics on the amino acid, e.g. tyrosine has an aromatic side chain that possesses an –OH group which can be phosphorylated
- Amino acids form a zwitterion at neutral pH (i.e. in the cytosol) with a positive charge on the amine group and a negative charge on the carboxyl group
- They act as buffers (see Figure 1.1 for titration curve)
- Excess dietary amino acids are broken down:
    i. amino portion is used to form urea in mammals
    ii. carbon skeletons are recycled for the formation of glucose via **gluconeogenesis (glucogenic)**, or of **ketone bodies (ketogenic)**
    ➔ *see 5.8 Amino acid breakdown and synthesis (p. 156) for metabolism of amino acids*

Property	Amino acid	Abbreviation		Side chain	Notes
		Three letter	One letter		
Aliphatic non-polar	alanine	ala	A	CH_3	
	cysteine	cys	C	$SHCH_2$	sulphur containing—can form disulphide bonds
	isoleucine	ile	I	$C_2H_5(CH_3)CH$	
	leucine	leu	L	$(CH_3)_2CHCH_2$	
	glycine	gly	G	H	smallest amino acid—makes up ~1/3 collagen
	methionine	met	M	$CH_3SCH_2CH_2$	initiator amino acid for all proteins; sulphur containing
	valine	val	V	$(CH_3)_2CH$	
Aliphatic polar—hydroxyl	serine	ser	S	CH_2OH	can be phosphorylated or glycosylated (O-linked)
	threonine	thr	T	CH_3CHOH	can be phosphorylated or glycosylated (O-linked)
Aliphatic polar—amide	asparagine	asn	N	$CONH_2CH_2$	can be glycosylated—N-linked
	glutamine	gln	Q	$CONH_2CH_2CH_2$	can be glycosylated—N-linked
Aromatic	phenylalanine	phe	F	$C_6H_5CH_2$	
	tryptophan	trp	W	$C_8H_5NHCH_2$	disrupts alpha helices
	tyrosine	tyr	Y	$C_6H_4OHCH_2$	polar—can be phosphorylated
Basic	arginine	arg	R	$(CN_3H_5)^+(CH_2)_3$	can be acetylated or methylated
	histidine	his	H	$C_3N_2H_3CH_2$	pK_a ~7
	lysine	lys	K	$(NH_3)^+(CH_2)_4$	can be acetylated or methylated
Acidic	aspartic acid	asp	D	COO^-HCH_2	normally negative
	glutamic acid	glu	E	$COO^-HCH_2CH_2$	normally negative
Cyclic	proline	pro	P	$C_3H_6NH_2^+$	disrupts alpha-helices; can be hydroxylated, e.g. in collagen

Table 1.2 Amino acid side chains R in R–CH(NH$_2$)COOH

Proteins

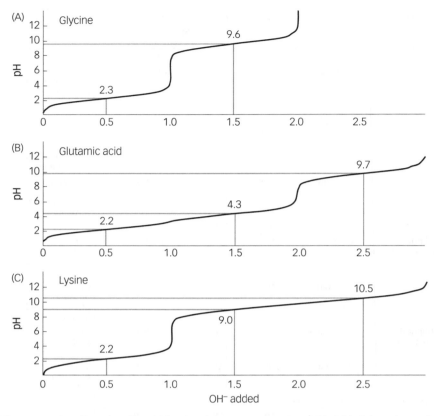

Figure 1.1 Titration of amino acids. Titration curves of amino acids with (A) uncharged, (B) acidic and (C) basic side chains. Grey line = midpoint of titration.

Key features of protein structure

- Four main levels of structure—primary, secondary, tertiary, and quaternary (see below)
- Three-dimensional shape is dependent on primary amino acid sequence
- Proteins fold to adopt the most thermodynamically stable conformation
- Chaperone molecules are required in the cell to stabilize partially folded intermediates and promote correct protein folding
- Proteins can show some flexibility in structure allowing conformational changes to regulate activity, e.g. allosteric regulator allolactose binds to lac repressor protein and leads to it dissociation from DNA

  ➜ *see Chapter 4: Regulation of transcription in prokaryotes (p. 94) for regulation of lac operon*
- Interior of the protein will contain hydrophobic interactions between side chains and ionic and H-bonds between polar/charged side chains

continued

- Exterior surface of the protein in contact with the aqueous environment must be hydrophilic—made up of charged or polar side chains to interact with water
- Exterior surfaces in contact with non-aqueous environments such as membranes contain amino acids with non-polar, hydrophobic side chains, but must be folded in the correct environment, e.g. in association with the membrane
- Three-dimensional structure of a protein is very precise and can be altered by a single amino acid change
- Precise structure of the active site of an enzyme is defined by the amino acids surrounding it and will define the range of substrates bound—any mutation that changes a single amino acid can lead to loss of enzyme activity
- **The precise shape of a protein is vital for correct function as it defines the interaction with other proteins, substrates, and regulatory molecules**

Primary structure of proteins

- Linear arrangement of amino acids
- Determined by order of nucleotides in mRNA (and ultimately DNA)
- Amino acids are linked together by **peptide bonds**

- Peptide bond (dotted box) has partial double bond character due to delocalization of electrons, so is *rigid* and *planar*
- R side chains are in the *trans* configuration, i.e. on opposite sides of the peptide bond
- All polypeptide chains have an amino [N] and carboxy [C] terminus
- Exist as zwitterions at physiological pH, i.e. at least one positive ($-NH_3^+$) and one negative ($-COO^-$) charge per molecule, plus further charges associated with the R side chains
- Depending on the number of acidic and basic side chains, the pH at which each protein is neutral will differ. This pH is known as the **isoelectric point** (pI). pH gradients can be used to separate proteins of different charge.
  - ➔ *see 9.9 Protein purification (p. 249) and 9.10 Protein analysis (p. 253) for protein separation*

Secondary structure of proteins

- Local folding of amino acids in three dimensions, defined mainly by H-bonds between residues of the peptide backbone (Figure 1.2)
- Two major forms of secondary structure:
  - i. **alpha-helix**
  - ii. **beta-pleated sheet**
- α-helix—hydrogen bonds every four residues within a strand. R groups face outwards
- β-pleated sheet—held together by hydrogen bonds between strands, either parallel or antiparallel

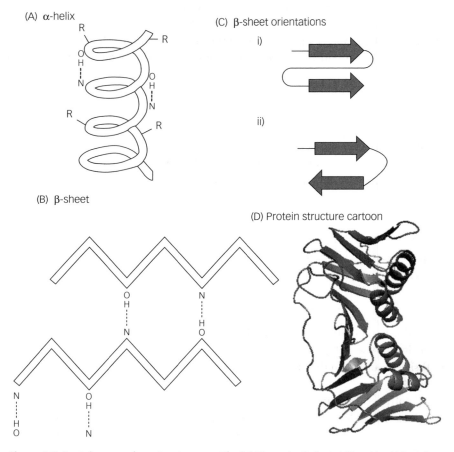

Figure 1.2 Protein secondary structure motifs. (A) The α-helix is stabilized by H-bonds within the chain. There are around 3.6 residues for every turn and the R groups face outwards. (B) The β-sheet is stabilized by H-bonds between strands. (C) In cartoons of protein structures, β-sheets are usually shown as thick arrows, which can lie (i) parallel or (ii) antiparallel to each other. The R groups alternately lie above and below the plane of the sheet. (D) Cartoon of a protein structure (monomer of PCNA) to show β-sheets and α-helices. Thin lines represent regions that are less ordered.

- Secondary structure of proteins is vital to their function, e.g. keratin (skin) is largely α-helix
- Collagen is made up of a triple helix (coiled coil) with glycine every third residue—small side chain (hydrogen) allows strands to come into close proximity for tight packing, important for structural strength. NB The collagen triple helix is unique and distinct from the α-helix

Tertiary structure of proteins

- Folding of the protein in three dimensions
- May involve interactions between residues that are far apart in the primary sequence
- Defined via interactions between the R side chains
- Tertiary structure stabilized by large number of mainly weak bonds—allows proteins to be flexible and change shape (important, for example, in allosteric regulation)
- Forces/bonding can be hydrophobic, ionic, or covalent (disulphide S–S bridges)
- S–S bonds form between cysteine residues which are brought together by tertiary structure and stabilize it—prevalent in proteins on cell surface or secreted from cell
- Tertiary structure commonly subdivided into **domains**, which are independently folding regions often with particular functions
- Most enzymes are globular proteins
- Globular proteins, e.g. myoglobin—ability to bind oxygen is defined by the precise tertiary structure (hyperbolic O_2 dissociation curve)
- Fibrous proteins serve structural roles, e.g. α-helical coiled coil of keratins in skin aids protein stability and confers strength
- Some proteins have both fibrous and globular regions, e.g. myosin in muscle—the structural region is fibrous (twisted α-helices) and the enzymatic ATPase activity is located in a globular head domain

Quaternary structure of proteins

- Arrangement of polypeptide chains in a multi-subunit protein, or inclusion of **prosthetic groups**
- e.g. haemoglobin: interaction of four subunits leads to co-operative binding of oxygen (**sigmoidal** dissociation curve for O_2 compared with hyperbolic for myoglobin, which has very similar unit structure)

Key evidence for protein structure

- Denatured *RNaseA* (no enzyme activity) can be refolded *in vitro* to restore activity—active structure is thermodynamically the most stable
- Single amino acid changes can alter protein function, e.g. mutant haemoglobin in sickle cell anaemia and thalassaemias
- Structures of many proteins and domains are known, and side chain interactions follow expected rules

- Structural evidence from:
    i. nuclear magnetic resonance (NMR) can be used to deduce the structure of polypeptides up to 200 amino acids long in solution
    ii. X-ray crystallography can be used to deduce the structure of larger proteins as long as high quality crystals are available
    iii. single particle electron microscopy (EM) and cryo EM, atomic force microscopy
    ➔ *see also 9.12 Biophysical techniques (p. 262) for methods used to determine protein structure*

1.3 LIPIDS

Lipids are water-insoluble, structurally diverse organic compounds. Functions include structural components of cell membranes, insulators (thermal, chemical, and electrical), energy stores, and biological signalling molecules.

Key features of lipids

- Most common **lipids** comprise **fatty acids** covalently bound to alcohol, e.g. glycerol or sphingosine; many variations in structure
- Two major types of lipids:
    i. hydrophobic (non-polar), e.g. **triglycerides**
    ii. **amphipathic** (polar head and non-polar tail), e.g. **phospholipids**
- Phospholipid polar head oriented toward the solvent and non-polar tails away from the solvent in aqueous solution—important in biological membranes
- Non-polar lipids are insoluble in aqueous solvents
- Amphipathic lipids form micelles (unstable) or self-healing bilayers (stable)
- Complex lipids form the structural basis of biological membranes and include:
    i. phospholipids (lipid + phosphate)
    ii. **sphingolipids** (lipid + sphingosine)
    iii. glycosphingolipids (sphingolipid + carbohydrate)
- Triglycerides (fats) are major molecules of fuel storage:
    i. fats are stored predominantly in **adipocytes** in mammals
    ii. stored fats metabolized to **acetyl CoA** by β oxidation in mitochondria, or formed into ketone bodies for utilization as preferred fuel of mammalian heart muscle, and of brain on starvation
    ➔ *see 5.6 Lipid breakdown and synthesis (p. 147) and 5.7 Ketone body breakdown and synthesis (p. 154) for lipid and ketone body metabolism*
    iii. carried as **lipoproteins** in mammalian blood

Fatty acids (FA)

- 12–20 carbons in aliphatic, unbranched hydrocarbon chain (usually even number) terminating in a carboxyl group
- Two major types
    - i. **saturated**, i.e. no double bonds
    - ii. **unsaturated**, i.e. contains C=C double bonds
- Unsaturated lipids can be:
    - i. monounsaturated has one double bond
    - ii. polyunsaturated has ≥2 double bonds
- Each fatty acid has specific T_m (melting temperature) dependent on how closely the individual molecules pack together
- Fatty acids are bound to serum albumin in blood for transport to heart, skeletal muscle, liver, etc. in mammals

Key aspects of fatty acid structure

- Fatty acids are often represented numerically as $x:y\Delta^z$, where:

    x = number of C in hydrocarbon chain

    y = number of double bonds

    z = position of double bonds (counted from carboxylate end)
- e.g. palmitate 16:0, oleate $18:1\Delta^9$, arachidonate $20:4\Delta^{5,8,11,14}$
- Carbons of fatty acid are numbered from the carboxyl carbon (C1); alternatively, first carbon of the hydrocarbon chain can be denoted the α carbon, and the last carbon in chain as ω (irrespective of chain length)
- Position of double bonds may be indicated with respect to ω carbon, e.g. oleate $18:1\Delta^9 = \omega\text{-}9$
- Double bonds in fatty acids either *cis* or *trans*

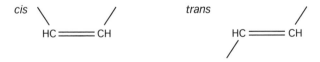

- *Cis* are more common in fatty acids found in nature
- *Cis* double bonds introduce bends or kinks into the hydrocarbon chain and lead to a decrease in T_m because the fatty acid chains cannot pack as closely
- *Trans* double bonds do not appreciably affect the direction of fatty acid chains— can pack tightly

Key physical properties of fatty acids

- Oxidation yields ~37 kJ/g
- T_m increases with length hydrocarbon chain; also degree of saturation
- Saturated fats are waxy at room temperature (e.g. butter)
- Unsaturated fats are generally liquid at room temperature (e.g. olive oil)

Lipids

- Forces: covalent, hydrophobic, and van der Waals

Triacylglycerols (triglycerides, TAG)

- Three fatty acid chains linked to glycerol (three-carbon alcohol = triol) via ester bond (condensation reaction)
- NB glycerol forms the backbone of many naturally occurring lipids
- Most abundant lipids in mammals
- Act as compact, neutral, anhydrous storage form of fatty acids
- Found especially as fat droplets in, for example, adipocytes of mammals
- Not present in biological membranes
- Major dietary source of lipids for mammals and major source of stored energy

glycerol   triglyceride

Glycerophospholipids (phosphoglycerides)

- Most abundant lipids in biological membranes
- Amphipathic (polar head, non-polar tail)
- Highly diverse class according to head group and/or two possible different fatty acid chains
- e.g. human red blood cells contain ≥21 different types of phosphatidylcholine which differ not in head group but in combination of fatty acid chains
- Generalized **phosphoglyceride**:

polar head   non-polar tail

Key types of glycerophospholipid

name	–O–X of general formula
phosphatidate	–O–H
phosphatidylcholine (PC)	$-O-CH_2CH_2N^+(CH_3)_3$
phosphatidylethanolamine (PE)	$-O-CH_2CH_2NH_3^+$
phosphatidylserine (PS)	$-O-CH_2-CH(NH_3^+)COO^-$

Ether phospholipids

- Formed by reduction of carbonyl groups of phospholipids
- Can serve as important signalling molecules, acting at low concentrations
- e.g. platelet activating factor (1-alkyl, 2-acetyl ether analogue of phosphatidyl choline)
    - i. promotes platelet aggregation
    - ii. promotes dilation of blood vessels

Sphingolipids

- Found in both plant and animal cell membranes, especially in mammalian central nervous system
- Sphingosine forms the core backbone with different fatty acids attached
- Sphingosine = C18 unbranched alcohol
- Ceramide is the metabolic precursor in the formation of all sphingolipids, with a fatty acyl group amide bonded to the C2 of sphingosine
- Three major classes of sphingolipids
    - i. sphingomyelins have phosphate group therefore = phosphosphingolipid
    - ii. cerebrosides have carbohydrate therefore = glycosphingolipids
    - iii. gangliosides have carbohydrate therefore = glycosphingolipids

Isoprenoids

- Include steroids
- Key example: **cholesterol** (found in animal plasma membranes, rarely in plants and never in prokaryotes)
- Other examples:
    - i. stigmasterol (plants)
    - ii. mammalian bile salts
    - iii. mammalian **hormones** such as the estrogens and androgens (e.g. testosterone)
    - iv. lipid vitamins, e.g. A, D, E, and K
    - v. sterols of plants, fungi, and yeast

Waxes

- Non-polar esters of long chain fatty acids and long chain monohydric alcohols
- Protective role, forming a waterproof coating on:
  - i. leaves and fruits of some plants
  - ii. skin or fur (mammals)
  - iii. feathers (birds)
  - iv. exoskeleton (insects)
- e.g. beeswax = ester of palmitate (16:0) and 30C myricyl alcohol

Eicosanoids

- Oxygenated derivatives of C20 polyunsaturated fatty acids, e.g. arachidonic acid
- Both physiological and pathological roles, e.g. vasoconstriction, inflammation, etc.
- Prostaglandins are eicosanoids with a cyclopentane ring

Enzymes that cleave phospholipids

- **Phospholipases** cleave phospholipids at specific positions
- Different classes according to position cleaved in phospholipid
- Digestive enzymes, e.g.
  - i. mammalian pancreatic PLA_2
  - ii. snake venom PLA_2
  - iii. secreted by bacteria, e.g. *Clostridium perfringens* α-toxin = *PLC*
- Important in signalling cascades, e.g. inflammatory cascade: PLA_2 generates arachidonic acid, a precursor of prostaglandins

1.4 CARBOHYDRATES

Carbohydrates make up most of the organic matter on earth. Carbohydrates are aldehydes or ketones with multiple hydroxyl groups or their polymers. They act as energy stores and fuels, as well as having a structural role.

Key features of sugars

- Hydroxylated ketones or aldehydes that can be linked by **glycosidic bonds** to form oligo- and polysaccharides
- Monomeric sugars act as intermediates in metabolism, e.g. glucose and other molecules in the glycolytic pathway

continued

➔ *see 5.2 Glucose breakdown and synthesis (p. 131) and 5.5. Glycogen breakdown and synthesis (p. 143) for carbohydrate metabolism*

- Dimeric sugars are broken down to monosaccharides for entry into key metabolic pathways

- Polysaccharides serve as compact energy storage molecules, e.g. glycogen and starch, or as structural elements, e.g. in plant cell walls (cellulose) and insect exoskeletons (chitin)

- Glycoproteins and glycolipids are found on cell surfaces and are probably involved in cell recognition

- Phosphorylated sugars form the structural framework of DNA and RNA

Monosaccharides

- Aldehydes and ketones with two or more hydroxyl groups $(CH_2O)_n$ where n ≥ 2 (see Table 1.3)
- Contain multiple asymmetric carbon atoms
- Naturally occurring hexoses tend to be D isomers e.g. D-glucose and D-fructose
- Pentoses and hexoses can cyclize to form ring structures
- Rings can be six-membered (pyranose) or five-membered (furanose)
- Can be linked together to form oligo- and polysaccharides
- Each ring structure can exist as either α or β forms:
  - ◦ α (alpha) sugars have the OH group of C1 below the plane of the ring
  - ◦ β (beta) sugars have the OH group of C1 above the plane of the ring
- The C1 carbon is the one with the aldehyde group on for aldoses, or the one adjacent to the ketone group of ketoses
- Glucose conformations:

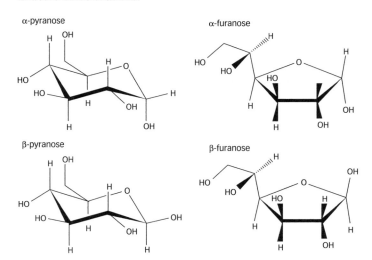

Carbohydrates

General name	Number of carbons	Type	Example
Triose	3	aldose	glyceraldehyde
	3	ketose	dihydroxyacetone
Hexose	6	aldose	glucose
	6	ketose	fructose

Table 1.3 Common sugars

Key monosaccharides

- Glucose (C_6 aldose)
- Galactose (C_6 aldose)
- Mannose (C_6 aldose)
- Fructose (C_6 ketose)
- *N*-Acetylglucosamine (C_6 aldose)
- *N*-Acetylgalactosamine (C_6 aldose)

Disaccharides

- Two monosaccharides joined together by **glycosidic bonds**
- These bonds can be either α or β configuration
- Bond formation requires activation of the sugar monomer in order to make it reactive, usually by transient association with a nucleotide triphosphate, e.g. UTP

Key disaccharides

Sucrose = glucose + fructose (α-D-glucopyranosyl-(1-2)-β-D-fructofuranoside)

Maltose = glucose + glucose (α-D-glucopyranosyl-(1-4)-α-D-glucopyranose)

Lactose = galactose + glucose (β-D-galactopyranosyl-(1-4)-α-D-glucopyranose)

Polysaccharides

- Multiple monosaccharide units linked together by glycosidic bonds to form a polymer
- Can serve as fuel stores, e.g. **glycogen** in animals and **starch** in plants
- Can have structural role, e.g. in plant cell wall (cellulose) and insect exoskeleton (chitin)

Glycogen

- Large branched polymer of glucose
- Most glucose units are linked by α-1,4 linkages; branches are formed by α-1,6 glycosidic bonds
- Branches form at approximately every 10th residue
- Fuel store in animal cells—predominantly liver and muscle of mammals

α-1,6 glycosidic link

α-1,4 glycosidic link

Starch

- Major fuel store in plant cells
- Very long polymers of glucose synthesized in chloroplast
- Can be stored in cytoplasm or chloroplast
- Exists with two components:
    - i. amylose—unbranched chains of glucose with α-1,4 linkages
    - ii. amylopectin—like amylose but with a branch with an α-1,6 linkage approximately every 30 residues

300-600

Dextran

- Fuel storage polysaccharide in yeast and bacteria
- Polymer of glucose
- Almost exclusively linked α-1,6, though a few branches with α-1,2, α-1,3, or α-1,4 linkages

Carbohydrates

α-1,4
α-1,6

α-1,6

Cellulose

- Structural molecule of plant cell walls, giving strength and stability
- Unbranched polymer of glucose with β-1,4 linkages
- Stabilizing H-bonds between adjacent glucose units give cellulose its tensile strength—not found in α linkage form (glycogen, starch)

Chitin

- Structural molecule of insects, forming exoskeleton
- Polymer of *N*-acetylglucosamine residues in β-1,4 linkage (like cellulose except for acetylamino group on C2)

Glycoconjugates

- Oligosaccharides can be attached to protein and lipids by glycosidic bonds
- Glycoprotein normally on extracellular domain:
    i. *O*-glycosidic bonds to serine or threonine
    ii. *N*-glycosidic bond to asparagine
- e.g. cell adhesion molecules (CAMs) are glycoproteins
- Sugars provide essential 'address label' on many proteins, e.g. lysosomal enzymes carry characteristic mannose 6-phosphate tag
- Sugar groups may also be important in the mechanics of protein trafficking within the cell, e.g. nuclear pore complex proteins are glycosylated, and nuclear import is blocked by lectins that bind to the sugars
- Glycolipids are found predominantly in plasma membranes of cells in the vertebrate nervous system—may be important both in cell recognition and in electrical insulation

Looking for extra marks?

Intracellular glycoproteins may be involved in protein recognition and targeting, e.g. nuclear pore complex proteins are glycosylated.

1.5 DNA

DNA (deoxyribonucleic acid) is the genetic material of most living organisms. It exists as a very long filamentous double helix (i.e. two strands of DNA).

Key features of DNA (deoxyribonucleic acid)

- Polymer of deoxyribonucleotides linked by **phosphodiester bonds**
- Chemically, **DNA** is made up of four nitrogenous bases (A, T, G, C) linked to a deoxyribose sugar bearing a phosphate group
- Phosphodiester bonds form between adjacent sugars
- **DNA double helix** is formed by complementary base pairing through hydrogen bond formation between bases on opposite strands of the double helix
- A always pairs with T, and G always pairs with C: the two strands are **complementary**
- Sugar phosphate backbone, which is negatively charged, interacts with aqueous environment
- Genetic code is determined by the order of the four nitrogenous bases

Chemical composition of DNA

- There are four **nitrogenous bases** in DNA: **cytosine** (C), **guanine** (G), **thymine** (T), and **adenine** (A)
- **Pyrimidines** (C and T) have a single nitrogenous ring:

- **Purines** (G and A) have double ring structure:

- **Deoxyribonucleosides** = base + sugar
    i. nitrogenous bases linked to the C1 carbon of 2′ deoxyribose by a β-*N*-glycosidic bond
- **Deoxyribonucleotides** = base + sugar + phosphate
    i. nucleoside with the 5′ carbon of deoxyribose bound by an ester linkage to a mono-, di-, or tri-phosphate group (dNMP, dNDP, or dNTP)
    ii. dNTPs are substrates for DNA synthesis during DNA replication
    ➔ *see 4.1 DNA replication (p. 78)*
    iii. general structure of a deoxyribonucleotide (dNTP):

Key structural features of DNA

- DNA is a polymer of four different deoxyribonucleotides (dAMP, dTMP, dCMP, dGMP) linked by phosphodiester bonds
    ➔ *see Chapter 4: DNA polymerases (p. 80) for bond formation*

- Polymer has polarity: the 5′ terminal nucleotide is a deoxyribonucleotide triphosphate; the 3′ terminal nucleotide has a free 3′ hydroxyl group—always drawn 5′→3′ by convention
- Acidic due to the phosphate groups
- Bases are hydrophobic and reside on inside of helix: sugar–phosphate backbone is hydrophilic and faces the aqueous environment
- Order of nitrogenous bases along the DNA determines the genetic code
- Two single strands of DNA coil around each other in anti-parallel fashion forming double stranded helix, or **duplex** DNA (see Figure 1.3), held together by **complementary base pairing**: A always base pairs with T (two hydrogen bonds); C always base pairs with G (three hydrogen bonds)
- Width of AT pair = width of GC pair (precise, so no helical disruption)
- DNA is a stable molecule but subject to spontaneous and environmental damage
  ➔ *see 4.4 DNA repair (p. 112) for details*
- Three major conformations of duplex DNA: A, B, and Z:
    i. B form DNA is hydrated right-handed helix, probably the form found in nature (see Figure 1.3)
        a. double helix with major and minor groove; helix diameter is 23.7 Å (2.37 nm)
        b. ten bases per turn of helix—3.4 Å (0.34 nm) per base giving a periodicity (one complete turn of the helix) of 34 Å or 3.4 nm
        c. hydrogen bonds between complementary base pairs are perpendicular to the axis of the helix
    ii. A is dehydrated right-handed helix, more tightly wound than B form
    iii. Z is left-handed helix, with no grooves, and occurs on repetitive alternating stretches of Pu/Py
- Bonding in double helix:
    i. hydrogen bonds between complementary base pairs: A forms two hydrogen bonds with T, and C forms three hydrogen bonds with G
    ii. hydrophobic effects (bases inside backbone)
    iii. base stacking interactions (van der Waals contacts—individually weak but additive, so significant for long DNA polymer)
    iv. charge–charge interactions: the large negative charge of the backbone that would otherwise destabilize DNA is generally neutralized *in vivo* by presence of cations (e.g. Mg^{2+}) and cationic proteins (e.g. histones in eukaryotes)

Key evidence for DNA structure

- Base pairing deduced from base ratios: A + G = T + C; A = T, G = C (Chargaff rules—from TLC analysis)
- X-ray crystallography (Franklin and Wilkins): diffraction pattern shows double helix

DNA

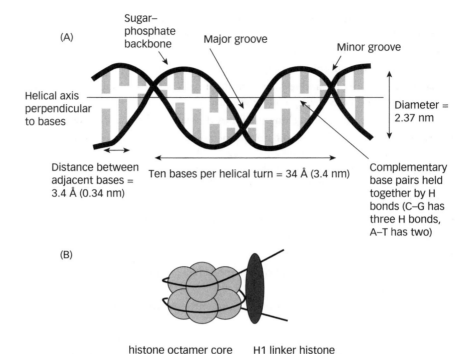

Figure 1.3 (A) DNA double helix. Two antiparallel strands of DNA, held together by H-bonds between the bases, twist to form a double helix with the hydrophilic sugar-phosphate backbone on the outside. There are ten bases per turn, aligned perpendicular to the helical axis. Note major and minor grooves. (B) Nucleosome. An octamer of core histones (two each of H2A, H2B, H3, and H4, pale grey) has two turns of DNA wrapped around it (approx. 146 bp). The linker histone H1 (dark grey) may be present at the entry/exit sites.

- Model building based on X-ray crystallography (Watson and Crick): bases must be on inside and purine must base pair with pyrimidine
- Thermal denaturation with A_{260} nm measurements give T_m (melting temperature) for DNA—absorbance increases sharply on disruption of cooperative interactions between stacked H-bonded base pairs

Higher order structure of DNA

Due to the long length of DNA compared with the size of cell, DNA must be compacted but remain accessible for replication and gene expression. A chromosome is a single molecule of double-stranded DNA. Prokaryotes have circular chromosomes and eukaryotes have linear chromosomes.

Prokaryotes

- e.g. *Escherichia coli* 4.2×10^6 bp of genomic DNA in one circular **chromosome** = 1.3 mm long; cell diameter <1 μm

- Circular DNA can be supercoiled (linking number = number of supercoils)
- May be bound by HU proteins which are basic
- Cell may also contain extra-chromosomal plasmids—usually large supercoiled circular DNA molecules (e.g. up to 100 kb) that may encode, for example, fertility factors or antibiotic resistance. NB Naturally occurring plasmids are much bigger than plasmids used for cloning in molecular biology (which may be ~3–10 kb)
- Genomic DNA of prokaryotes is located in the cytoplasm as **nucleoid**
- In prokaryotes, much of the DNA encodes proteins or structural RNA molecules

Eukaryotes
- e.g. *Homo sapiens* ~2.9×10^9 bp of DNA in 2×23 chromosomes = 1.8 m long, 2 nm diameter; nuclear diameter <10 μm
- Compartmentalized into the **nucleus** of eukaryotes
- Complex of DNA with associated proteins is known as **chromatin**
- DNA packaged into hierarchical structure giving overall 10^4-fold compaction
    i.   duplex DNA 2 nm diameter
    ii. wrapped around octamer of basic histone proteins (core nucleosome), giving 'beads on a string' appearance in EM (11 nm fibre)
    iii. 30 nm diameter fibre with linker histone H1
    iv. 300 nm diameter solenoids
    v.   1,400 nm diameter metaphase bivalent chromosomes (each sister chromatid 700 nm diameter)
- Nucleosome consists of a histone octamer (two each of four core histones H2A, H2B, H3, and H4) with two turns of DNA wrapped around (146 base pairs) (see Figure 1.3B)
- Linker distance between nucleosomes can vary
- Actively transcribed genes are found in more loosely packed chromatin known as **euchromatin**
- Regions of DNA not containing actively transcribed genes are often more tightly packed into **heterochromatin**
- Euchromatin and heterochromatin are associated with different histone modifications

    ➔ *see Chapter 4: Regulation of initiation of transcription in eukaryotes (p. 99) for histone modifications associated with gene expression*

Key evidence for higher order structure
- DNA spreads from bacterial cells or eukaryotic chromosomes treated with detergent and/or EDTA; examination by electron microscopy (EM) (size of genomic DNA relative to cell) shows large amount of DNA spilling out from lysed cell
- X-ray crystallography (nucleosome)

DNA

- EM shows various levels of packaging
- Light microscopy (chromosomes)
- Partial micrococcal nuclease digestion of eukaryotic chromatin—cuts DNA to produce a ladder of fragments of approx. 140 bp as the DNA is cleaved in linker DNA between regularly spaced nucleosomes
- Agarose gels ± treatment with topoisomerase to separate differently supercoiled forms demonstrates supercoiling in prokaryotes

Key functions of DNA

- DNA is the genetic material of the cell
- Coding information is arranged as genes (~1–10% of higher eukaryotic DNA is made up of protein coding genes) and intergenic regions, together with specific structural and regulatory regions
- Majority of genomic DNA in prokaryotes and lower eukaryotes codes for proteins or structural RNA molecules
- In higher eukaryotes such as man, ~1% of the genome codes for proteins

Genes

- Encode structural proteins (via the intermediate mRNA) and functional RNA molecules (rRNA, tRNA, snRNA, microRNA, and other regulatory RNAs)
- Order of the bases in the genes determines the order of amino acids in proteins
  ➔ *see 4.3 Protein synthesis (translation) (p. 101)*
- Eukaryotic coding regions of genes (**exons**) can be interrupted by non-coding **introns** (intervening sequences)—rare in yeast, also found in some bacteria and bacteriophage

Intergenic regions (between genes) in eukaryotes include

- Promoters and gene regulatory sequences
  ➔ *see 4.2 RNA synthesis (transcription) (p. 88)*
- Origins of replication
  ➔ *see 4.1 DNA replication (p. 78)*
- Matrix attachment sites to anchor DNA to proteinaceous structure in nucleus
- **Telomeres** (ends of chromosomes)—important for chromosomal stability; measure of cell age as they shorten in each round of DNA replication in somatic cells
- **Centromeres** (middles of chromosomes)—essential for attachment to microtubules of spindle during cell division to ensure accurate segregation of sister chromatids or homologous chromosomes to daughter cells in mitosis or meiosis, respectively
- Repetitive DNA elements, e.g.
    i. microsatellite and minisatellite, e.g. Alu in *H. sapiens*
    ii. transposable elements (e.g. copia) and interspersed elements (LINES, SINES)

- Uncharacterized regions—originally called 'junk DNA'? but much of this DNA may encode critical regulatory small RNAs

Key evidence that DNA is the genetic material

- White blood cells from pus on surgical bandages treated with HCl gave a precipitate containing carbon, hydrogen, oxygen, nitrogen, and phosphorus – called 'nuclein' (Miescher 1869)
- Factor could be transferred from heat-killed pathogenic (smooth) pneumococci to non-pathogenic (rough) strain—the 'transforming principle' (Griffith 1928) (Figure 1.4)
- Transforming principle was resistant to lipases, proteases, and heat, but was destroyed by treatment with nucleases. Purification showed that the principle was DNA (Avery, MacLeod, and McCarty 1944)
- Bacteriophage radiolabelled with ^{35}S (proteins) or ^{32}P (nucleic acid) infected into bacteria. Only the ^{32}P transferred into the bacterial cell, therefore genetic material of phage is nucleic acid not protein (Hershey and Chase 1952) (Figure 1.5)
- DNA and protein sequencing combined with mutational analysis to alter DNA bases in bacteria demonstrates that order of bases in DNA defines order of amino acids in protein
- Classical genetics—mutations in DNA leading to changes in phenotype
- Genome sequencing projects including the EST (expressed sequence tag) database and specific projects to sequence the DNA from individual species (e.g. Human Genome Project) relate DNA sequence to product
- Fluorescence *in situ* hybridization (FISH) demonstrates presence and location of specific stretches of DNA on chromosomes, e.g. telomere-specific probes
- Density gradient ultracentrifugation can be used to purify repetitive elements of particular density, demonstrating that they exist as repeats rather than as single copies

Exam tip

In questions about DNA structure, make sure you draw a clear, correctly labelled diagram of DNA—see Figure 1.3.

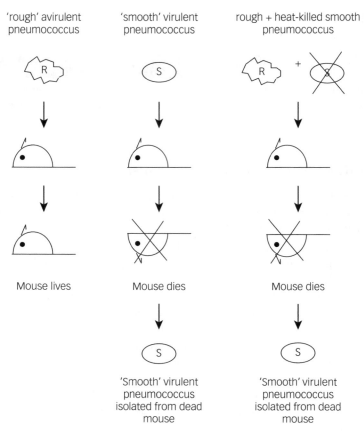

'rough' avirulent pneumococcus

'smooth' virulent pneumococcus

rough + heat-killed smooth pneumococcus

Mouse lives

Mouse dies

Mouse dies

'Smooth' virulent pneumococcus isolated from dead mouse

'Smooth' virulent pneumococcus isolated from dead mouse

Figure 1.4 Transforming principle (Avery). Mice were infected with one of two strains of pneumococcus bacteria: The rough 'R' form is non-lethal but the smooth 'S' strain kills the mice. It is possible to heat-kill the 'S' bacteria (proteins are denatured, DNA remains intact); these dead bacteria do not cause illness in mice. However, if a mouse is infected with both the heat-treated S form and the non-pathogenic R form, the R strain of the bacteria can take up DNA from the dead S strain and acquire their pathogenic properties, and so the mice die. The bacteria isolated from the dead mouse are now the virulent 'S' strain.

1.6 RNA

RNA (ribonucleic acid) is a single-stranded polymer of ribonucleotides linked by phosphodiester bonds. Some viruses have an RNA genome (mammalian retroviruses, e.g. human immunodeficiency virus). RNA can have structural, regulatory, and catalytic roles as well as acting as the intermediate between DNA and proteins. RNA is made during the process of transcription, and is complementary to the sequence of the DNA from which it is transcribed.

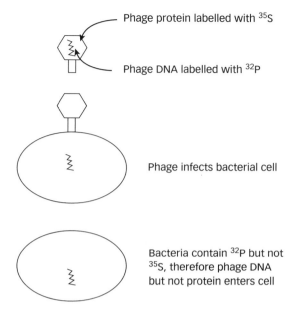

Phage protein labelled with ^{35}S

Phage DNA labelled with ^{32}P

Phage infects bacterial cell

Bacteria contain ^{32}P but not ^{35}S, therefore phage DNA but not protein enters cell

Figure 1.5 Genetic material is nucleic acid (Hershey and Chase). Bacteriophages were radioactively labelled with ^{35}S in the protein and ^{32}P in the nucleic acid. Only the ^{32}P entered the cell and was incorporated into new phage particles. Therefore the genetic material is nucleic acid, not protein.

Chemical composition of RNA

- **Ribonucleosides** = base + sugar, i.e. nitrogenous bases linked to the C1 carbon of ribose by a β–*N*-glycosidic bond (note *deoxy*ribose in DNA)
- **Ribonucleotides** = base + sugar + phosphate, i.e. nucleoside with the 5′ carbon of ribose bound by an ester linkage to one or more phosphate groups. rNTPs act as substrates for **RNA** synthesis during transcription
  - ➜ *see 4.2 RNA synthesis (transcription) (p. 88)*
- General structure of a ribonucleotide (rNTP, aka NTP):

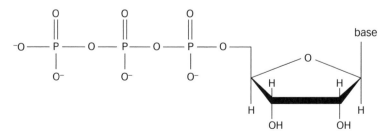

- Bases in RNA:
  - i. pyrimidines (C and U) have a single nitrogenous ring
  - ii. purines (G and A) have double nitrogenous ring structure
  - ➔ *(see 1.5 DNA (p. 20))*
  - iii. structure of **uracil (U)**:

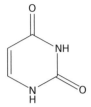

Key features of ribonucleic acid (RNA)

- Polymer of ribonucleotides linked by phosphodiester bonds
- RNA is composed of four different types of nitrogenous base linked by a glycosidic bond to the C1 of ribose (compared to deoxyribose in DNA)
  - i. sugar in RNA has OH at 2′ position as well as 3′ position seen in DNA—makes RNA more chemically reactive than DNA
  - ii. nitrogenous bases of RNA: adenine (A), cytosine (C), guanine (G), and uracil (U) (instead of thymine in DNA)
- Polymer has polarity, with a triphosphate on the 5′ carbon of the ribose at one end and a free OH group on the 3′ carbon of the other terminal nucleotide—always drawn 5′→3′ by convention
- RNAs are single-stranded but in physiological solution can adopt various secondary structures according to sequence
- Palindromic regions of complementary base pairs can form hairpin loops, e.g. stem–loop structure of tRNA molecules and termination loop during transcription.
- Double-stranded RNA (e.g. some viruses) resembles 'A form' double-stranded DNA

Key classes of RNA

- rRNA
  - ○ ribosomal RNA
  - ○ functions as structural and enzymatic component of ribosomes
  - ○ 80% of total cellular RNA
- mRNA
  - ○ messenger RNA
  - ○ acts as intermediate between genetic code of DNA and amino acid sequence of proteins

- ◦ 3% of total cellular RNA
- tRNA
  - ◦ transfer RNA
  - ◦ adapter molecule that carries amino acids to specific codon on mRNA in ribosome during protein synthesis
  - ◦ 73–95 nucleotides long
  - ◦ 5% total cellular RNA
- snRNA
  - ◦ small nuclear RNA, in eukaryotes
  - ◦ involved in removal or splicing of introns from eukaryotic mRNA
  - ◦ catalytic or associated with catalytic proteins
- miRNA
  - ◦ small RNA (~22 nucleotides), often complementary to untranslated region of mRNA
  - ◦ involved in regulating gene expression
  - ◦ ~1,000 in human genome, regulating ~60% of genes
- piRNA
  - ◦ piwi-interacting RNA
  - ◦ 26–31 nucleotides long
  - ◦ largest class of mammalian small RNAs
  - ◦ role in transposon and gene silencing
- siRNA
  - ◦ short interfering double-stranded RNA molecules
  - ◦ usually 21 bases long with an overhang of two at each 3′ terminus
  - ◦ recruits RNA-induced silencing complex (RISC) for mRNA cleavage and gene silencing
  - ◦ used experimentally to knock down specific gene expression

 Check your understanding

Compare the chemical nature and structure of DNA and RNA. (*Hint: remember RNA uses uracil instead of thymidine; the extra OH group at the 2′ position in RNA increases its reactivity; higher order structures due to base pairing.*)

How does the primary sequence of a protein define its final structure? (*Hint: remember all four levels of protein structure and the different bonding types involved; give examples of globular and structural proteins as well as membrane proteins.*)

Describe the structure of three polymers found in cells. (*Hint: make sure you chose polymers that you can draw and describe in adequate detail.*)

What is the structure of glucose and its naturally occurring polymers? (*Hint: remember to include starch and glycogen.*)

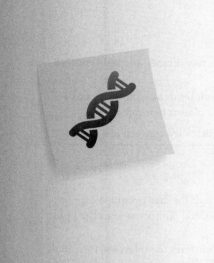

2 Cellular components

2.1 MEMBRANES

Membranes are made up of lipid, protein, and carbohydrate, and form a hydrophobic barrier to allow compartmentalization. The plasma membrane separates the contents of a cell from its environment. Eukaryotes have intracellular compartments defined by membranes that allow specialization of certain regions within the cell to increase the efficiency of biochemical processes.

Key features of biological membranes

- Fluid mosaic model of membrane structure (see Figure 2.1)
- **Amphipathic** phospholipids form a **bilayer** with the polar heads facing the hydrophilic exterior of cell or aqueous cytoplasm
- Fatty acid tails form a hydrophobic interior
- Sheets of bilayers form spontaneously in aqueous environments, are held together by non-covalent interactions, and are self-healing
- Proteins are embedded in this 'sea' of phospholipids (**integral**) or associated with the surface (**peripheral**)
- Protein interaction with phospholipids is non-covalent, e.g. hydrophobic amino acid side chains can interact with the fatty acid tails in the centre of the bilayer, charged ones with the polar lipid head groups

continued

- Protein content varies greatly between different membranes; the lipid content varies, but less dramatically
- Carbohydrates can be covalently attached to both lipids and proteins, and are found mainly on the extracellular surface of the plasma membrane
- Huge variety of carbohydrate modifications which vary between cell types

Key evidence for membrane structure

- Surface area of a red blood cell is approximately half the area occupied by extracted lipids in a monolayer on an aqueous surface, consistent with the lipids forming a bilayer in cell membrane
- Electron density/X-ray across a membrane demonstrates the bilayer nature, with electron density suggestive of the phospholipid head groups at either edge and a central hydrophobic region twice the length of the average fatty acyl chain
- Some proteins can be removed from membranes by high salt, i.e. they are peripheral

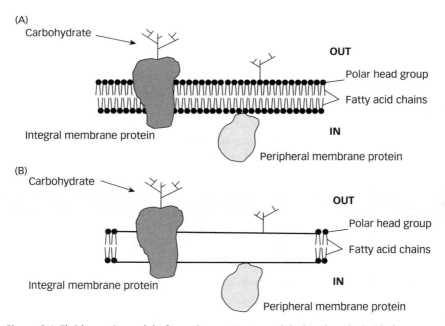

Figure 2.1 Fluid mosaic model of membrane structure. (A) The phospholipids form a bilayer with polar head groups facing outwards and the hydrophobic fatty acid chains buried on the inside. Proteins can span this fluid bilayer (integral) or be associated with one face (peripheral). Complex branched carbohydrates are attached to some proteins and lipids. These are most abundant on the outside of the plasma membrane. (B) Note that the accurate way of drawing membranes takes far too long in exams—it is perfectly valid to simplify to that shown in B.

- Other proteins can only be extracted with organic solvents/detergents which disrupt the lipid bilayer, i.e. they are integral
- Freeze–etch electron micrographs show more proteins associated with the internal face and fewer with the external, i.e. not all proteins span the lipid bilayer.
- Some proteins can be labelled in intact cells and so must be exposed on outside face of membrane. After cell permeabilization (without bilayer disruption) to allow reagent access to inner surface of membrane, the same proteins can be labelled more extensively and so they must span the membrane
- Lectins, which bind carbohydrates, bind only to the extracellular face of the plasma membrane, not the intracellular face

Membrane fluidity

- Membranes are fluid structures with both lipids and proteins diffusing laterally
- Movement of a lipid from one half of the bilayer to the other (flip-flop) is very rare except when facilitated by enzymes, e.g. during the biosynthesis of lipids in the endoplasmic reticulum (ER)
- Rotation of a protein across the membrane has not been observed: membrane asymmetry is therefore preserved
- Fluidity allows integral membrane proteins to change conformation without compromising the permeability barrier, e.g. in regulating growth factor receptors or protein channels
- Membrane fluidity is controlled by:
  i. length of fatty acyl chain and number of double bonds (especially in prokaryotes)
  ii. proportion of cholesterol (eukaryotes)—inserts into bilayer and alters phase transition of fatty acyl chains
- All lipids are mobile and diffuse rapidly: some proteins diffuse as fast as lipids, others are virtually immobile
- Lipid rafts = regions of the membrane rich in atypical lipids: important in regulating membrane fluidity, protein, and receptor trafficking and signalling

Looking for extra marks?

Proteins that don't move are anchored by association with the extracellular matrix on the outside of the cell and/or the actin cytoskeleton on the inside.

Key evidence for membrane fluidity

- Proteins on the surface of two cells can be labelled with dyes which fluoresce at different wavelengths. When the two cells are fused, the dyes become intermingled, demonstrating that the proteins are free to move in the plane of the bilayer (see Figure 2.2)

Membranes

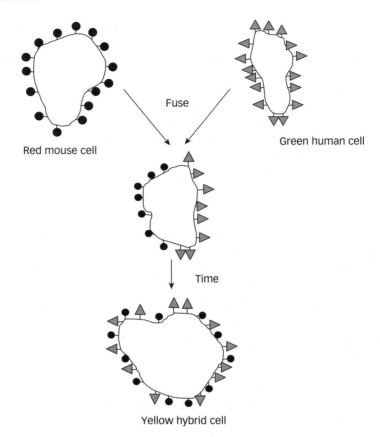

Fuse

Red mouse cell

Green human cell

Time

Yellow hybrid cell

Figure 2.2 Evidence for fluid membranes. A mouse cell is labelled with a membrane-binding red dye and a human cell with a green membrane-binding dye. After fusion the dyes intermingle to produce a cell that fluoresces yellow. Dyes can be attached to lipids or most proteins to show they both move. Some proteins don't show mobility—these must be anchored within the membrane. (Exam tip: use coloured pens to draw each membrane—do NOT try to draw the symbols.)

- Fluorescence recovery after photobleaching (FRAP) demonstrates the mobility of lipids. Lipids are fluorescently labelled and a small patch is bleached using a laser. The patch 'disappears' as labelled lipids exchange with the bleached lipids (see Figure 2.3)
- Electron spin resonance (ESR) of spin-labelled lipids. Lipids are labelled on one side only and remain on that face of the bilayer, demonstrating no flip-flop
- Measuring phase transition temperature T_m of lipid bilayers with differing fatty acyl chains and cholesterol content demonstrates different fluidity
- The decrease in degree of saturation of fatty acyl chains from thigh to foot in reindeers is consistent with this as a mechanism of long-term control of fluidity
- Different fatty acyl content of *E. coli* grown at high and low temperatures demonstrates physiological adaptation of fluidity properties

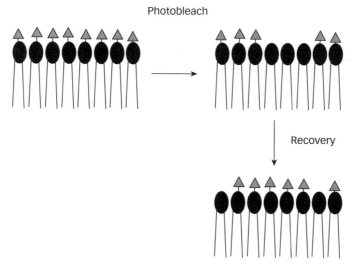

Photobleach

Recovery

Figure 2.3 FRAP—fluorescence recovery after photobleaching. Phospholipids can be labelled with a fluorescent dye. A laser is used to bleach a spot on the cell, which now appears dark in the fluorescence microscope. Over time, labelled phospholipids diffuse into the spot. The rate of recovery of fluorescence reveals how quickly the lipids are moving. It is also possible to fluorescently label membrane proteins to discover how quickly they move in the membrane.

Functions of membrane components

1. Lipid

- Permeability barrier—permeable only to:
    - i. small polar molecules (e.g. gases such as O_2, CO_2)
    - ii. small hydrophobic molecules (e.g. fatty acids, steroid hormones)
    - iii. water: high molarity of water coupled with the high concentration of solutes inside the cell means that water can cross the lipid bilayer, despite the hydrophobic interior of membranes
- Solvent for integral membrane proteins
- Specific modification of some lipids can be a mechanism of transmitting an extracellular signal into the cell, e.g. *phospholipase C* cleavage of inositol lipids in vasopressin signalling

    ➔ *see 6.5 Principles of hormone signalling in metabolism (p. 183) for details on vasopressin signalling*

Looking for extra marks?

A small proportion of lipids interact with specific membrane proteins—necessary for catalytic function of protein, e.g. protein kinase C.

2. Proteins

- Mediate many membrane functions
- Carriers of, for example, glucose (**GLUT transporters**) to allow selective permeability
- Form gated channels to allow ion flow, e.g. in neurones
- Receptors for hormones (e.g. insulin) and other signals (e.g. growth factors)
- Cell adhesion to adjacent cells or extracellular matrix—usually glycosylated
- Enzymes to catalyse membrane-based reactions, e.g. *receptor tyrosine kinases*

3. Carbohydrate

- Important in cell–cell recognition, e.g. essential for sperm/egg interaction
- Can act as 'clocks' or timers—sialic acid is modified and removed from surface over time, thus distinguishing old cells from new, e.g. lymphocytes
- Role in cell adhesion, e.g. lymphocyte binding to sites of tissue damage

2.2 CYTOSKELETON

The cytoskeleton is composed of networks of different types of filaments (microfilaments, intermediate filaments, and microtubules) which are present throughout the cell. They serve as communication lines for cellular traffic, to allow cell movement, to stabilize cell structure, and to protect from physical stress.

Key components of cytoskeleton

Three main types of cytoskeletal elements, named according to diameter in electron micrographs (decreasing size below):

  i. **Microtubules** composed of tubulin are involved in cell motility (cilia and eukaryotic flagella) and in intracellular transport, e.g. transfer of vesicles from ER to Golgi apparatus, and spindle in cell division
  ii. **Intermediate filaments** form stable structures with high tensile strength, e.g. keratins of skin
  iii. **Microfilaments** composed of actin are involved in cell motility (e.g. amoebae), intracellular transport, and in contractile processes (e.g. muscle contraction and cytokinesis)

Microtubules

- Polymers of α and β tubulin dimers form microtubules 25 nm in diameter
- Microtubules are hollow cylinders of 13 protofilaments
- *In vitro*, given an excess of free tubulin, each protofilament (and microtubule) has polarity with a plus (rapidly growing) and minus (slowly growing) end. When the concentration of free tubulin drops below the critical concentration, the rate of depolymerization will be faster than the rate of **polymerization** (see Figure 2.4)

- In the cell, microtubules are anchored at the equivalent of the 'minus' end and grow/retract from the 'plus' end—filaments are polarized
- Centrosomes are the primary site at which the minus ends of microtubules are anchored— also called microtubule organizing centres (MTOCs)
- Microtubules grow towards periphery of the interphase cell and are continually polymerizing and depolymerizing—dynamic instability
- Capping proteins stabilize ends
- Tubulin is a GTP binding protein:
  i. GTP hydrolysis not necessary for microtubule polymerization but can explain the dynamic instability of microtubules as the hydrolysis of GTP to GDP induces a conformational change which favours depolymerization
  ii. if addition of subunits outstrips GTP hydrolysis microtubule is growing rapidly; if addition of subunits outstrips GTP hydrolysis GTP-bound cap will protect end from depolymerization
  iii. this cap is lost as the rate of polymerization slows
  iv. when the end subunits are all in the GDP-bound form, the rate of depolymerization increases

(A) Microtubule organisation in the cell

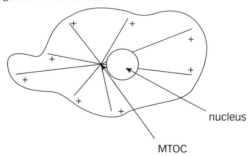

nucleus

MTOC

(B) Dynamic instability of microtubules.

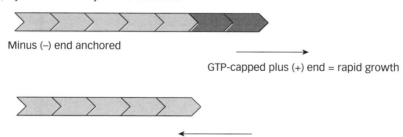

Minus (–) end anchored

GTP-capped plus (+) end = rapid growth

Loss of GTP-capped end = rapid shrinkage

Figure 2.4 Microtubules. A. Microtubules grow from a centrosome (also known as a microtubule organizing centre or MTOC) near the nucleus, with the 'plus' ends to the edge of the cell. B. The tubulin dimers polymerize in a GTP-bound form. A cap of GTP-bound tubulin promotes rapid polymerization. Loss of the GTP-bound 'cap' causes rapid depolymerization. Microtubules show dynamic instability.

Cytoskeleton

Functions of microtubules

- Structural and transport roles in the cell
- Directed movement of organelles and vesicles using motor proteins (kinesin and dynein), e.g. in neurones
- Spindle formation and chromosome segregation during cell division
- Cilia and flagella in eukaryotes (NB: a different protein—flagellin—makes up bacterial flagella)

Evidence of microtubule assembly and roles

- Immunostaining of microtubules using fluorescently labelled anti-tubulin antibodies reveals their location in different cell types and changes in organization as the cells go through a round of cell division
- Polymerization of microtubules *in vitro* demonstrates polarity of microtubule growth, role of GTP, and its hydrolysis as well as dynamic instability
- In presence of non-hydrolysable GTP analogues, polymerization occurs normally but microtubules show reduced stability, demonstrating the role of GTP hydrolysis in microtubule stability
- Disruption of microtubules by drugs blocks mitosis and disrupts vesicle transport:
    i. destabilizing drugs, e.g. nocodazole
    ii. stabilizing drugs, e.g. taxol

Looking for extra marks?

Drugs that disrupt tubulin assembly or disassembly are used to treat some cancers as they block cell division (e.g. taxol).

Intermediate filaments

Intermediate filaments are polymers of coiled coil subunits, e.g. keratins. They are responsible for protection from mechanical stress.

- Filaments of 8–10 nm in diameter
- Monomers of many types all with same basic structure
- Central rod regions of two monomers form α-helical coiled coil, i.e. dimer
- Two anti-parallel dimers produce a tetramer
- Soluble tetramers form fundamental units which polymerize into a helical array to form a filament (see Figure 2.5)
- Components of filaments vary greatly with cell type, e.g. keratins in skin, desmin in muscle
- Form a dense network throughout the cytoplasm
- Specific nuclear intermediate filaments (lamins) line the nucleoplasmic face of the nuclear envelope—they are dynamic and disassemble prior to nuclear envelope breakdown in mitosis

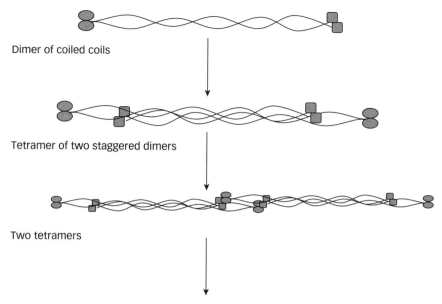

Dimer of coiled coils

Tetramer of two staggered dimers

Two tetramers

Eight tetramers twist into a rope-like filament

Figure 2.5 Intermediate filaments. Dimers of subunits form a coiled coil with globular termini. Tetramers form of two staggered dimers and ropes of eight tetramers twist to form filaments.

Functions of intermediate filaments

- Resistance to mechanical stress
- Rigid support for otherwise flexible membrane structures

Evidence for intermediate filaments

- Polymerization *in vitro* demonstrates biochemical requirements and stoichiometry
- Human blistering disease caused by mutation in keratin (epidermolysis bullosa simplex) demonstrates that lack of keratins renders cells very sensitive to mechanical injury
- Mechanical properties of the filaments *in vitro* is consistent with a role in protection from stress

Actin microfilaments

Microfilaments are polymers of actin that form a dense network under the cell surface. They are responsible for defining cell shape, for attachment of cells to their substrate, and are important in cellular movement.

- Polymer of uniformly oriented actin monomers forming two intertwined helices = microfilament
- 8 nm in diameter

Cytoskeleton

- Called F-actin (filamentous actin) when polymerized
- Globular actin (G-actin) can self-assemble *in vitro* to form actin filaments with polarity:
    i. plus/barbed (rapidly growing polymerization) end
    ii. minus (slowly growing/depolymerization) end
- Actin is an ATP binding protein: ATP-bound form involved in assembly
- ATP hydrolysis is not necessary for filament formation but plays a role in filament turnover, analogous to that played by GTP bound to tubulin
- *In vitro*, when the critical concentration of free actin is reached, actin filaments show **treadmilling** where the rate of polymerization at the plus end equals the rate of depolymerization at the minus end (Figure 2.6)
- In the cell, actin polymerization is regulated by a large number of interacting proteins which can cap ends, promote polymerization or depolymerization, or sever actin filaments. Pattern of polymerization controlled by small G proteins (e.g. rac, rho)
- Actin filaments form a mesh under the surface of the cell
- In muscle cells, actin filaments are organized into parallel rows to allow muscle contraction on association with myosin
- Actin filaments have specific attachment points at the cell membrane where the cell adheres to adjacent cells or extracellular matrix: stress fibres
- Growth of actin filaments is responsible for amoeboid cell movement by creating the pseudopodia at the cell's leading edge

Functions of microfilaments

- Cell adhesion by interacting with integral membrane proteins such as integrins
- Motility: chemotaxis, cytokinesis, and muscle contraction (with myosin)
- Determine membrane shape, e.g. intestinal microvilli
- Maintain cell polarity

Evidence for microfilaments

- Polymerization *in vitro* demonstrates ATP dependence, polarity of growth, and treadmilling
- In presence of non-hydrolysable ATP analogues, polymerization occurs normally allowing the function of ATP hydrolysis to be separated from binding. ATP hydrolysis is necessary to depolymerize filament
- Staining of actin filaments using fluorescently labelled antibodies/phalloidin demonstrates the location of the filaments in different cell types and in response to extracellular signals
- Disruption of actin filaments in cells using drugs such as cytochalasin D demonstrates those processes for which actin filaments are required
- Striations seen in microscopy of skeletal muscle represent bands of actin and myosin

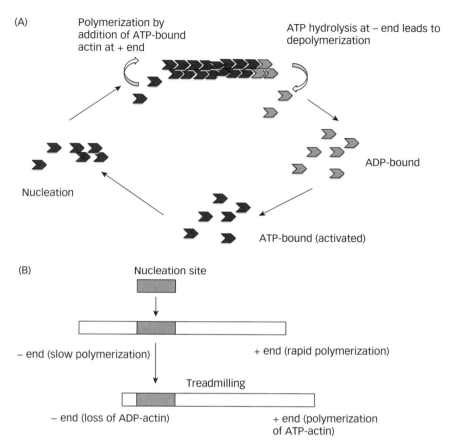

(A)

Polymerization by addition of ATP-bound actin at + end

ATP hydrolysis at – end leads to depolymerization

ADP-bound

Nucleation

ATP-bound (activated)

(B)

Nucleation site

– end (slow polymerization)

+ end (rapid polymerization)

Treadmilling

– end (loss of ADP-actin)

+ end (polymerization of ATP-actin)

Figure 2.6 Actin polymerization. Polymerization requires nucleation leading to rapid growth of the 'plus' end and slower growth of the 'minus' end. A. Schematic representation of stages: ADP-bound globular actin does not polymerize; ATP binding leads to activation; nucleation of activated actin monomers to form oligomers; polymerization into twisted filaments—addition at growing end; ATP hydrolysis to ADP at minus (–) end; and dissociation of ADP-bound actin from minus (–) end. B. Simplified diagram suitable for exams. Treadmilling at steady state, where loss of ADP-actin from 'minus' end equals rate of polymerization of ATP-actin at 'plus' end.

2.3 NUCLEOSKELETON

The nucleoskeleton is the structure underlying the organization of genomic DNA within the nucleus. Interphase chromatin is attached via specific DNA sequences to a skeleton (also known as the nuclear scaffold or matrix). The nature of the skeleton is still controversial but may include scaffolding proteins plus RNA. The scaffold may attach to the nuclear envelope via the nuclear lamina.

Nucleoskeleton

Nuclear matrix

- Fibrous structure present when majority of DNA is removed
- Contains scaffold proteins, e.g. SCI = *topoisomerase II*
- DNA attached via scaffold (or matrix) attachment regions (SARs or MARs), which are usually AT-rich sequences

Functions of nuclear matrix

- May cluster functional sites of DNA, e.g. transcription promoters, DNA replication origins, etc. in small volume of nucleus
- Reduces 'dimensionality' problem of very large enzyme complexes locating sites of action on DNA from three dimensions to only one dimension (a punctate 'spot')
- May stabilize fragile interphase chromatin by attachment to nuclear lamina and internal support structures
- Permits final organization of chromosomes

Evidence for nuclear matrix

- *DNase 1* treatment of nuclei which removes >90% DNA leaves functional replication and transcription regions which can be labelled by endogenous *DNA* and *RNA polymerases* incorporating fluorescent dNTPs or NTPs respectively
- 2 M salt treatment of nuclei followed by electron microscopy reveals dense filamentous structures (though may be artefact of high salt used)

Nuclear lamina

- Nuclear lamina is made up of a specialized form of intermediate filament proteins called lamins
- Central rod domain is longer than in most intermediate filament proteins
- Lamins can assemble into a two-dimensional lattice which requires other proteins
- Lattice is dynamic as it disassembles during cell division (controlled by phosphorylation)

Functions of nuclear lamina

- Protects nuclear contents
- Stabilizes the nuclear membrane to form a barrier between the nucleus and cytoplasm during interphase of cell cycle
- May assist in nuclear envelope reformation from vesicles after mitosis

Evidence for nuclear lamina

- Immunodepletion of lamins blocks the reformation of the nuclear envelope following mitosis

- Electron microscopy reveals network of lamin filaments in nucleus underlying the nuclear envelope
- Mutation in lamin A gene causes severe destabilization of nuclear structure (and fatal childhood premature ageing Hutchinson–Gilford progeria)

2.4 CYTOSOL

The cytosol is the aqueous material remaining after extraction of insoluble components (nucleus, membranes, organelles, and cytoskeleton) by ultracentrifugation.

Key features of the cytosol

- Aqueous
- Approximately neutral pH (~pH 7.2) and physiological [salt], e.g. ~150 mM in *Homo sapiens*, 74 mM in *Xenopus laevis*
- Prepared by ultracentrifugation of lysed cells which removes membrane systems and polymerized cytoskeletal elements
- Contains soluble enzymes and enzyme complexes plus protein 'factories', e.g. ribosomes
- Storage site of glycogen (animals)
- Topologically distinct from lumens of membrane-bounded organelles

Functions of cytosol

- Site of many enzyme-catalysed reactions, e.g. **glycolysis**
  - *see 5.2 Glucose breakdown and synthesis (p. 131) for details of glycolysis*

Evidence for cytosol

- Ultracentrifugation—glycolysis enzymes remain in soluble fraction
- Conductivity measurements reveal salt concentration of **cytosol**

 Check your understanding

Describe the structure of microtubules. What is their function? (*Hint: remember that microtubules are important in defining cell shape (e.g. neurones), as well as in formation of the mitotic spindle.*)

What makes actin filaments dynamic? (*Hint: remember the importance of ATP hydrolysis and the polarity of the filaments. Mention why they need to be dynamic—in cell motility and when cells round up for division; actin is involved in cytokinesis in association with myosin, as well as its better known interaction with myosin in muscles.*)

2.5 ORGANELLES

Eukaryotic cells are larger than prokaryotic cells, and are compartmentalized into membrane-bounded organelles. Animal cells (Figure 2.7) and plant cells (Figure 2.8) share major similarities in organelles, but have some structural and organizational differences.

Key concepts of organelles

- **Compartmentalization** improves efficiency of multistep reactions, as products from one reaction are usually co-localized with the enzyme catalysing the next step
- Reduces possibility of side reactions
- Allows different concentrations of metabolites in different regions, e.g. $[Ca^{2+}]$ in sarcoplasmic reticulum/cytoplasm, NAD^+/NADH ratios in cytoplasm/mitochondria
- Reduces amount of protein required, as proteins can be concentrated in specific functional regions of the cell, e.g. histones at high concentration in nucleus
- Facilitates cell specialization—relative size and number of organelles varies according to cell type, e.g.
    i. increased volume of endoplasmic reticulum in cells synthesizing steroid hormones
    ii. increased number of mitochondria in cells requiring large amounts of ATP such as muscle cells
- Delicate molecules can be protected from physical stresses, e.g. DNA in nucleus away from shearing forces in cytoplasm in motile or contractile cells
- Keeps potentially destructive molecules away from sites where they could do damage, e.g. nucleases and proteases in lysosomes
- Spatial separation of discrete processes allows greater regulation, e.g. transcription in nucleus, translation in cytoplasm in eukaryotes
- Prokaryotes lack membrane-bounded compartments but show some degree of compartmentalization, e.g. electron transport chain proteins are located to specific regions of the plasma membrane

Nucleus

The nucleus is the storage site of genomic DNA within a eukaryotic cell. DNA replication and transcription take place here, together with RNA processing. mRNA and other RNAs must be exported from the nucleus in order to function in, for example, translation. The nucleus is bounded by a double membrane (the nuclear envelope), studded with very large multiprotein nuclear pore complexes that regulate trafficking across the nuclear envelope.

Key features of nucleus

- Approximately 10 μm diameter (varies considerably) (see Figure 2.9)
- Contains genomic DNA packaged into chromatin

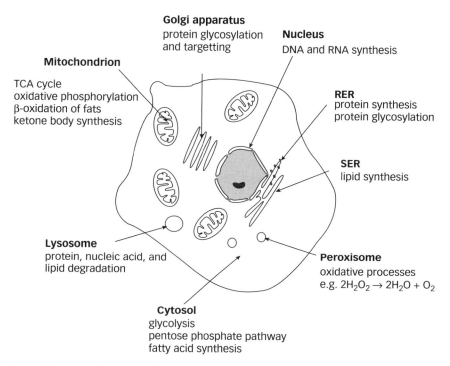

Golgi apparatus
protein glycosylation
and targetting

Nucleus
DNA and RNA synthesis

Mitochondrion

TCA cycle
oxidative phosphorylation
β-oxidation of fats
ketone body synthesis

RER
protein synthesis
protein glycosylation

SER
lipid synthesis

Lysosome
protein, nucleic acid, and
lipid degradation

Peroxisome
oxidative processes
e.g. $2H_2O_2 \rightarrow 2H_2O + O_2$

Cytosol
glycolysis
pentose phosphate pathway
fatty acid synthesis

Figure 2.7 A typical animal cell.

- Surrounded by nuclear envelope:
    - i. two phospholipid bilayers (outer and inner nuclear membranes)
    - ii. intermembrane space continuous with ER lumen
- Nuclear envelope punctuated with **nuclear pore complexes** (NPCs) that regulate transport into and out of nucleus:
    - i. very large—125×10^6 Da (cf. ribosome ~4×10^6)
    - ii. multiprotein assemblies composed of >100 different proteins
    - iii. eight-fold symmetry with 'basket' structure
    - iv. central aqueous channel permits free diffusion of many proteins <60 kDa
    - v. regulated opening to allow specific traffic in both directions of larger components via binding to carrier proteins (importins)
    - vi. specific import usually requires a nuclear localization sequence (NLS) whereas export requires a nuclear export sequence (NES)—these are short amino acid sequences within the protein that is being transported
- Meshwork nuclear lamina underlies nuclear envelope, providing structural stability
- Nuclear envelope breaks down into smaller vesicles during prophase and reforms in telophase of mitosis (cell division)
- DNA probably anchored on nuclear scaffold/matrix or nucleoskeleton via scaffold/matrix attachment sites on DNA bound by specific scaffold proteins, e.g. *topoisomerase II*

Organelles

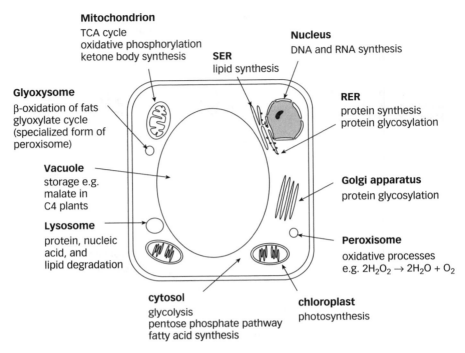

Mitochondrion
TCA cycle
oxidative phosphorylation
ketone body synthesis

Nucleus
DNA and RNA synthesis

SER
lipid synthesis

Glyoxysome
β-oxidation of fats
glyoxylate cycle
(specialized form of
peroxisome)

RER
protein synthesis
protein glycosylation

Vacuole
storage e.g.
malate in
C4 plants

Golgi apparatus
protein glycosylation

Lysosome
protein, nucleic
acid, and
lipid degradation

Peroxisome
oxidative processes
e.g. $2H_2O_2 \rightarrow 2H_2O + O_2$

cytosol
glycolysis
pentose phosphate pathway
fatty acid synthesis

chloroplast
photosynthesis

Figure 2.8 A typical plant cell.

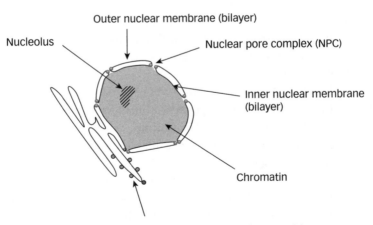

Outer nuclear membrane (bilayer)

Nucleolus

Nuclear pore complex (NPC)

Inner nuclear membrane
(bilayer)

Chromatin

Outer nuclear membrane is continuous with ER

Figure 2.9 Typical nucleus. The nucleus is bounded by a double membrane with
proteinaceous nuclear pores. The outer membrane is continuous with the ER, though
lipid and protein flow is probably blocked by the nuclear pore complexes. The chromatin is amorphous in appearance in interphase (between cell divisions) with the nucleolus
visible as an electron-dense structure (without a membrane) in electron microscopy.

- Nucleolus visible as darker region NOT bounded by separate membrane; contains high levels of *RNA polymerase I*, rRNA transcripts and assembling ribosomes

Functions of nucleus
- Site of DNA replication, transcription, and RNA processing
- Nucleolus = site of rRNA transcription and partial assembly of ribosomes
- Compartmentalization of RNA transcription in nucleus away from site of translation (in cytoplasm) allows for greater control of gene expression, e.g. splicing out of introns
- Protects long, fragile genomic DNA from shearing forces, e.g. during cell motility
- Allows concentration of enzymes and precursors necessary for DNA replication and transcription
- Nuclear scaffold may reduce 'dimensionality problem' by localizing replication enzymes at replication origins, and by clustering transcription start sites (promoters)

Evidence for nucleus
- Transmission electron microscopy (EM) shows double membrane of nuclear envelope and electron-dense nucleolus
- EM shows 'Christmas tree' structures in nucleolus—many molecules of *RNA polymerase I* transcribing tandem repeats of rDNA
- Three-dimensional reconstruction of EM frozen sections shows nuclear pore complexes in various states of opening
- Lectins (e.g. wheat germ agglutinin) bind to NPCs on outer nuclear envelope—can be used to purify NPCs or selectively block nuclear protein import
- 2 M salt extraction reveals scaffold (caveat: artefactual conditions)
- *DNase* and *RNase* treatment of nuclei embedded in agarose under physiological salt shows replication and transcription sites clustered and protected from digestion

Mitochondria

Mitochondria are organelles with a double membrane and serve as the site for the TCA cycle and oxidative phosphorylation, together with other metabolic processes such as β-oxidation of fats.

Key features of mitochondria
- Approximately 1.5–10 μm long × 0.25–1.0 μm diameter, though can exist as large reticular networks (Figure 2.10)
- Surrounded by double membrane (each of which is a bilayer)—outer and inner mitochondrial membranes (OMM and IMM, respectively)
- Matrix and intermembrane space are aqueous

Organelles

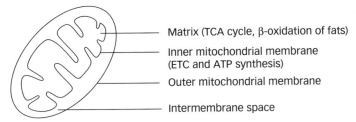

Matrix (TCA cycle, β-oxidation of fats)

Inner mitochondrial membrane
(ETC and ATP synthesis)

Outer mitochondrial membrane

Intermembrane space

Figure 2.10 The mitochondrion is bounded by a double membrane. The matrix is the site of the TCA cycle and oxidation of fatty acids. The inner membrane contains the complexes of the electron transport chain (ETC) and is the site of ATP synthesis.

- Inner membrane provides electrical and chemical insulation
- Outer membrane relatively freely permeable due to presence of porin
- Electron transport chain (ETC) proteins embedded in inner mitochondrial membrane
    - ➔ *see 5.4 Oxidative phosphorylation (p. 140) for electron transport chain*
- F_1/F_0 *ATP synthase* on inner mitochondrial membrane
- Regions where membranes juxtapose may represent sites for protein import—such proteins usually have an **amphipathic** α-helix which acts as a mitochondrial import signal
- Number of mitochondria may reflect metabolic requirements of cell
- Duplicated throughout the cell division cycle
- Equally partitioned between daughter cells on cell division
- Contain own DNA encoding some, but not all, subunits of essential ETC proteins (other subunits encoded by nuclear DNA), plus own rRNAs and tRNAs
- Transcribe own DNA and translate the resulting mRNA—ribosomes resemble those in prokaryotes
- **Endosymbiotic theory**—mitochondria may have arisen from microorganisms

Looking for extra marks?

In sexual reproduction in animals, mitochondria are inherited predominantly from the egg (i.e. cytoplasmic inheritance), so diseases associated with organelle DNA are inherited from the mother.

Functions of mitochondria

- Tricarboxylic acid (TCA, also known as citric acid or Krebs) cycle takes place in mitochondrial matrix
    - ➔ *see 5.3 Tricarboxylic acid (TCA) cycle (p. 137)*
- Electron transport chain generates electrochemical proton gradient across IMM
- Major site of ATP synthesis by oxidative phosphorylation—proton flow through *ATP synthase* down gradient coupled to ATP synthesis

- Generation of heat: uncoupling proteins allow flow of protons down electrochemical gradient without synthesis of ATP
- β-Oxidation of fats in matrix
  - ➜ *see 5.6 Lipids (p. 149) for details of fatty acid metabolism*
- Some reactions of urea cycle in liver mitochondria
  - ➜ *see Chapter 5: Urea cycle (p. 159)*

Evidence for mitochondria

- EM shows structure with double membrane
- Mitochondrial ribosomes are similar to those of bacteria—supports endosymbiotic theory

Chloroplasts

Chloroplasts are organelles with double membrane systems found only in plants and photosynthetic algae. They are the sites of photosynthesis.

Key features of chloroplasts

- Present only in plants and photosynthetic algae
- Site of photosynthesis
  - ➜ *see 8.1 Photosynthesis (p. 202) for details of photosynthesis*
- Approximately 10 μm long × 0.5 μm diameter (Figure 2.11)
- Bounded by double membrane:
  - i. highly permeable to CO_2
  - ii. selectively permeable to other metabolites
- Aqueous **stroma** contains soluble enzymes that catalyse **reduction** of CO_2 to carbohydrate during **photosynthesis**; triose phosphate = exported carbohydrate
- **Thylakoid membranes**
  - i. internal membrane stacks
  - ii. highly folded and continuous membrane network
  - iii. contain chlorophyll a and b

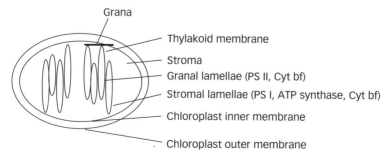

Figure 2.11 The chloroplast is the site of photosynthesis in plants. It is bounded by a double membrane but also contains membranous stacks or grana.

Organelles

- Accessory pigments (e.g. carotenoids and phycobilins) complement light absorption spectra of chlorophyll to permit light absorption from wide range of wavelengths (400–700 nm)
- **Grana** = flattened thylakoid membrane vesicles arranged in stacks
- **Stroma lamellae** are in contact with stroma and contain various enzyme complexes, e.g. **photosystem** I: *ATP synthase*
- **Grana lamellae** only in contact with other membranes of granum—contain other enzyme complexes, e.g. photosystem II (oxygen evolution): Cyt bf complex
- Contain own DNA encoding some components of photosystems (some variations in 'universal' genetic code). Other components are encoded by nuclear DNA and proteins are imported from cytosol
- Chloroplast ribosomes are distinct from those in cytoplasm and more like those of bacteria
- May have arisen from **symbiotic** photosynthetic bacteria in eukaryotic cells— endosymbiotic theory

Functions of chloroplasts

- Photosynthesis, i.e. CO_2 reduction to carbohydrate using water and energy from light

  ➔ *see 8.1 Photosynthesis (p. 202) for details of photosynthesis.*
- Starch storage and mobilization

Evidence for chloroplasts

- Electron microscopy for overall arrangement of membranes
- Extraction of different sub-compartments with biochemical analysis for enzymes (electrophoresis, immunology, functional **assays**)
- Reactions carried out by isolated chloroplasts *in vitro*

Endoplasmic reticulum (ER)

The endoplasmic reticulum is a very large reticular network bounded by a single membrane, serving as the site of some protein synthesis and biosynthesis of lipids.

Key features of the ER

- Single continuous membrane-bounded structure throughout much of cell
- ER membrane constitutes >50% total of cell membranes
- Lumen (cisternum) of ER may occupy >10% total cell volume
- Two types (Figure 2.12):
  - i. rough, with ribosomes (= RER), involved in synthesis of many proteins
  - ii. smooth, without ribosomes (= SER), involved in lipid synthesis
- Vesicularizes on mitosis for equal partitioning of components to each daughter cell

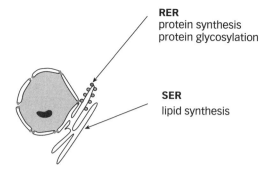

RER
protein synthesis
protein glycosylation

SER
lipid synthesis

Figure 2.12 The endoplasmic reticulum (ER) membrane is continuous with the outer membrane of the nucleus. The rough ER (RER) is studded with ribosomes and is the site of synthesis of proteins that will be inserted into membranes, subsequently targeted to the lumen of various membrane-bound vesicles via the Golgi, or secreted from the cell. The smooth ER (SER) is the site of lipid biosynthesis. Vesicles bud from the ER to take protein and lipid to the Golgi body.

Functions of ER
- ER membrane provides very large surface area for the transition between cellular compartments: the major site in the cell at which proteins change topological compartments, e.g. cytosol to lumen of organelle
- Proteins synthesized on ER ribosomes (rough ER) can be:
    i. inserted into membranes (during translation)
    ii. folded in the ER lumen as they are being translated, by chaperones in ER lumen, ready for targeting elsewhere
- Initial modification of newly translated proteins in ER lumen:
    i. cleavage of signal sequence
    ii. addition of ten-sugar complex
- Proteins are transferred to Golgi via vesicles budded from ER membrane
- Lipid biosynthesis (smooth ER)
- Acts as Ca^{2+} store (especially sarcoplasmic reticulum in muscle)

Evidence for ER
- EM—rough and smooth regions
- Studied using microsomes—vesicularized ER formed on maceration of tissue, e.g. liver
- SER and RER microsomes sediment separately on ultracentrifugation (mass of RER greater due to attached ribosomes)
- Yeast genetics—mutants with defective secretion have been used to identify protein components of ER membranes

Golgi apparatus (Golgi body)

The Golgi body is important in sorting proteins that were synthesized in the endoplasmic reticulum, attaching certain sugars, and in targeting the proteins for their eventual destination (e.g. lysosome). It is visualized in electron microscopy as a stack of elongated membrane-bound sacs usually near to the nucleus, with vesicles detectable between the sacs.

Key features of the Golgi body

- Series of membrane stacks (Figure 2.13) located in perinuclear region
- Vesicles transfer contents from *cis* through medial stacks to *trans* Golgi network (TGN)
- *Cis* stacks are adjacent to endoplasmic reticulum
- Prominent feature of cells specialized for secretion, e.g. goblet cells in intestine

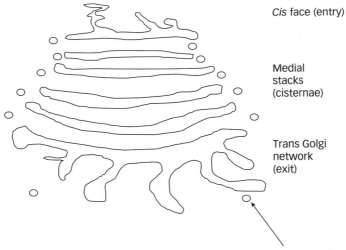

Cis face (entry)

Medial stacks (cisternae)

Trans Golgi network (exit)

Vesicle transported to other organelles (e.g. lysosome) or to plasma membrane

Figure 2.13 The Golgi apparatus or body is made up of a series of membranous stacks. Vesicles from the ER fuse with the *cis* stack, then vesicles carry lipids and proteins through successive stacks to the trans Golgi network (TGN). Proteins and lipids are glycosylated as they pass through and finally sorted in the TGN for different destinations, e.g. plasma membrane, secretory vesicles, lysosomes.

Functions of Golgi apparatus

- Modifies oligosaccharide chains present on proteins entering from ER
- Modifications are sequential as proteins pass from *cis* to medial to *trans* Golgi compartments
- Proteoglycan assembly by *O*-linked glycosylation of serine and threonine residues
- Proteins are sorted in the *trans* Golgi network and packaged into separate vesicles to be targeted to different locations in the cell, e.g. lysosomes (mannose-6-phosphate tag), plasma membrane, etc.

Evidence for Golgi apparatus

- Visualization in EM
- Localization of modifying enzymes to different stacks of Golgi by histochemical stains in EM, consistent with sequential modification
- Disruption of Golgi apparatus with Brefeldin A disrupts targeting of proteins to other membrane compartments

Lysosomes

Lysosomes are small vesicles that are important for degradation of both intracellular components and macromolecules taken up from outside the cell by endocytosis.

Key features of lysosomes

- Vesicles 0.05–0.5 μm in diameter
- Low intralysosomal pH (~pH 4–5) maintained by membrane-bound H^+-*ATPase*
- Heterogeneous population
- Up to 40 types of hydrolytic enzymes (*nucleases, proteases*, etc.)—all work optimally at acidic pH
- Lysosomal proteins are very destructive so must be targeted from Golgi directly to lysosome—lysosome-specific mannose-6-phosphate 'tag' on proteins is detected by a specific receptor in TGN

Function of lysosomes

- Intracellular digestion of cellular macromolecules, e.g. proteins, lipids, nucleic acids
- Fusion with endocytotic vesicles to digest components taken up from outside the cell

Evidence for lysosomes

- Raising pH with weak bases (e.g. ammonia) prevents function
- Labelling cells with pH-sensitive fluorescent probes reveals pH and heterogeneity in size

Peroxisomes

Peroxisomes are similar in size to lysosomes but function in oxidative metabolism, particularly in detoxification reactions.

Key features of peroxisomes

- Vesicles, approximately 0.5 μm in diameter
- Found in all eukaryotic cells
- Contain oxidative enzymes at very high levels
- Major site of oxygen utilization after mitochondria

Functions of peroxisomes

- Carry out oxidative reactions
- Detoxification, e.g. oxidation of alcohol, and other molecules
- Detoxify reactive oxygen species produced in oxidative reactions—contain, for example, *catalase* ($2H_2O_2 \rightarrow 2H_2O + O_2$) and *superoxide dismutase*
- Keep reactive oxygen species away from macromolecules that could be severely damaged by them (e.g. DNA)
- In plants and yeast, β-oxidation of fatty acids is exclusive to peroxisomes; in mammals this takes place in mitochondrial matrix
- Specialized form of peroxisomes in plants, **glyoxysomes**, carry out glyoxylate cycle to enable plants to convert fatty acids to carbohydrate (important in seed germination)

Evidence for peroxisomes

- Electron microscopy of vesicles coupled with histochemical staining for oxidative enzymes

Vacuoles

Vacuoles are membrane-bounded organelles of very variable size, usually important in storage.

Key features of vacuoles

- Present in plant and fungal cells
- Bounded by single membrane

- Aqueous contents
- Can occupy more than 90% of total cell volume, e.g. in mesophyll cells of some plants

Functions of vacuoles

- Storage of molecules that may otherwise inhibit enzymatic activities in the cytosol, e.g. [malate] ≤ 0.2 M in mesophyll vacuoles of some plants—very acidic
- Osmotically active—turgor pressure in plants

Evidence for vacuoles

- Light microscopy—indicates size of vacuole
- pH and [salt] measurements
- Electron microscopy—single membrane

 Check your understanding

Draw a typical eukaryotic animal cell and label the compartments. (*Hint: see Figure 2.7.*)

Which organelles have their own genetic information and how is this believed to have come about? (*Hint: see mitochondria and chloroplasts as well as nucleus; remember endosymbiotic theory.*)

Compare and contrast the components of the eukaryotic cytoskeleton (*Hint: don't forget intermediate filaments as well as microtubules and actin microfilaments.*)

What are the advantages of having intracellular compartments? (*Remember that bacteria lack compartmentalization, are specialized for rapid division and better at this than most eukaryotes, so there must be a strong reason for compartmentalization—think about different modes of metabolism and the risks of oxidative metabolism and degradative enzymes to other cellular components.*)

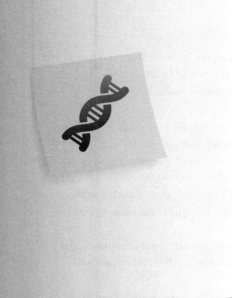

3 Enzymes

Enzymes are highly specific biological catalysts. Most are proteins though some RNAs can act enzymatically.

> ### Key concepts
>
> - Accelerate the rate of a reaction without altering the equilibrium position
> - Operate by stabilizing **transition states** for reactions
> - Participate in the reaction by making non-covalent or covalent bonds with intermediates in the reaction, but are regenerated unchanged at the end
> - Highly specific for individual substrate or group of related substrates, e.g. single compound, single **stereoisomer,** or related substances

3.1 COMMON FEATURES OF ENZYMES

- Mostly globular proteins, though a very few enzymes are made of RNA (*ribozymes*)
- Increase reaction rate by 10^3–10^{17} over uncatalysed reactions
- **Substrate** binds in **active site** in enzyme—often hydrophobic cleft
- Catalytically efficient—for some enzymes every collision between enzyme and substrate results in reaction

- Chemically efficient—high reaction specificity results in almost total conversion of substrate to **product** without formation of wasteful by-products
- Enzymes operate by stabilizing **transition states** for reactions
- Two models for action:
    i. **lock and key hypothesis**: enzyme = lock (preformed shape in active site) and transition state = key (fits specifically into active site)
    ii. **induced fit hypothesis**: active site models itself around transition state (by altering positions of amino acid side chains during binding)
- Common catalytic mechanisms involve **acid/base** catalysis and nucleophilic catalysis
- Large variety of enzymes catalyse vast majority of cellular reactions (≥ 1000 enzymes in simple micro-organisms)
- Different subsets of enzymes expressed in different cell types of multicellular organisms. Also different **isoenzymes** (same reaction, different regulation)
- Many enzymes operate reversibly in cells
- Those enzyme-catalysed reactions that are essentially irreversible form key regulatory points in metabolic pathways
- Regulated enzymes ('regulatory') are structurally more complex than unregulated enzymes, e.g. possess both catalytic and regulatory domains
- Enzymes have a range of functions but can be classified into six main groups (Table 3.1)

Class	Reaction catalysed	Notes
Oxidoreductases	redox	addition/removal of electrons, e.g. dehydrogenases
Transferases	group transfer	may need coenzyme, or enzyme may transitorily bind group
Hydrolases	**hydrolysis**	water acts as acceptor group
Lyases	non-hydrolytic and non-oxidative elimination to form double bond	reverse reaction = addition of one substrate to double bond of second substrate; enzymes now called *synthases*
Isomerases	structural change within a single molecule, i.e. **isomerization**	*isomerases, epimerases, mutases*
Ligases	joining (ligation) of two substrates	require energy from NTP; often called *synthetases*

Table 3.1 Six major classes of enzymes

3.2 ENZYME KINETICS

By measuring enzyme rates under different conditions (different substrate concentrations, pH, temperature, etc.), various aspects of their chemical mechanisms can be deduced. Such studies are generally termed enzyme kinetic studies.

Analysis of the rate of an enzyme-catalysed reaction

- Enzyme (E) forms transient complex with substrate (S), which then dissociates to form either enzyme plus product (P), or enzyme plus substrate:

$$E + S \leftrightarrow ES \leftrightarrow E + P$$

- At start of a reaction, the concentration of product is very low so reaction $ES \rightarrow E + P$ is essentially irreversible and $v = v_o$
- Rate of reaction depends on:
  - i. concentration of enzyme [E]
  - ii. concentration of substrate [S]
- v_0 is the rate of the enzyme-catalysed reaction at the origin of a progress curve, i.e.

$$v_0 = \Delta[P] / \Delta t \text{ at } t = 0 \; (\textbf{initial rate})$$

- **Progress curve** (see Figure 3.1): measures the rate of increase of product [P] or decrease of substrate [S] with time (t)
- The formation of an enzyme–substrate (ES) complex leads to saturation kinetics
- When E is **saturated** with substrate, reaction is essentially independent of [S] (*zero order with respect to [S]*) so only dependent on [E] (*pseudo first order*) (Figure 3.2)

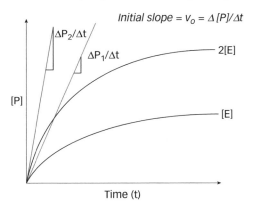

Figure 3.1 Progress curve: time course of product production in an enzyme-catalysed reaction.

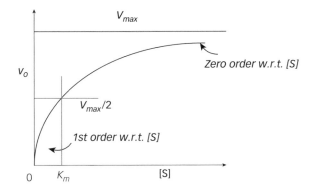

Figure 3.2 Dependence of enzyme rate on substrate concentration.

Enzyme kinetics

- Formation of ES complex usually involves formation of non-covalent bonds
- Conversion of ES $\rightarrow$ EP may be rate limiting because of chemical change that is undergone
- Alternatively, rate limiting step may be a conformational (tertiary structure) change in the enzyme
- Kinetics can be modelled mathematically on the basis of **rate constants** for elementary steps (Figure 3.3)

$$E + S \underset{k_{-1}}{\overset{k_1}{\rightleftharpoons}} ES \xrightarrow{k_{cat}} E + P$$

> k_1 is association rate constant for E + S
> k_{-1} is dissociation rate constant for ES
> k_{cat} is turnover number

Figure 3.3 Elementary steps of a simple enzyme-catalysed reaction.

Michaelis–Menten equation

$$v_o = \frac{V_{max}[S]}{K_m + [S]} = \frac{k_{cat}[E][S]}{K_m + [S]}$$

- Describes relationship between v_o and [S]
- Equation predicts enzyme reactions are first order with respect to [E] and show hyperbolic dependence on [S] (see Figure 3.2)
- Derivation of the Michaelis–Menten equation uses **steady state assumption**, i.e. [ES] is constant since the complex is formed as fast as it decomposes
- Derivation of the Michaelis–Menten equation assumes [S]>>[E]. This is normally true *in vitro*, but may not be true inside cells

k_{cat}

- Catalytic constant, k_{cat}, defined as maximum number of substrate molecules that can be converted to product per active site of enzyme per second
- Describes rate of enzyme turnover when [S] is saturating, e.g. *catalase* $k_{cat} = 10^7 \ s^{-1}$
- $V_{max} = k_{cat}[E]$—can only be derived from V_{max} if [E] is known
- k_{cat} is independent of [E], but V_{max} is proportional to [E]

K_m

- Michaelis constant is known as K_m
- $K_m = [S]$ when $v_0 = V_{max}/2$, i.e. when enzyme is half-saturated with substrate
- The *lower* the K_m, the *more tightly* the substrate is bound to enzyme
- K_m is independent of [E]
- K_m can be determined in impure preparations as long as only one enzyme in the preparation can catalyse the reaction measured
- Enzymes normally operate in cells with K_m near the concentration of the substrate

- $K_m \approx$ [S] in cells so as to allow the cell to respond proportionally to changes in [S]; helps to prevent substrate accumulation, cf. *glucokinase* in liver (see Table 3.2)

Enzyme	Organ	K_m	Role	Notes
Glucokinase	liver	high (~10^{-2} M)	converts glucose to glucose-6-phosphate for glycogen storage	not saturated at blood [glucose]
Hexokinases I, II, III	muscle etc.	low (~10^{-6} to 10^{-4} M)	phosphorylation of glucose at start of glycolysis	saturated at physiological [glucose]

Table 3.2 The isoenzymes of glucokinase/hexokinase serve different roles according to K_m

k_{cat}/K_m

- Equals apparent second order rate constant when [S] is non-saturating
- Measure of **substrate specificity** and **catalytic efficiency** of enzyme

Measuring K_m and V_{max}

- Enzyme activity is measured in an **enzyme assay**, by measuring appearance of product or disappearance of substrate
- Assays are performed at constant temperature and pH, at appropriate values to give high activity without appreciable denaturation of enzyme
- Measure K_m and V_{max} from plots of v_0 at a series of [S]
- Computer fitting data to the Michaelis–Menten hyperbola is most accurate way of determining constants
- Data can be linearized (easier for manual plotting) by converting the Michaelis–Menten equation into double reciprocal plot (Figure 3.4) using the Lineweaver–Burk equation

Lineweaver–Burk equation

$$\frac{1}{v_o} = \frac{K_m}{V_{max}} \frac{1}{[S]} + \frac{1}{V_{max}}$$

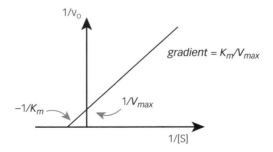

Figure 3.4 Lineweaver-Burk plot of rate/substrate relationship.

- $1/K_m$ is determined from the intercept of line with the x axis $(1/[S])$
- $1/V_{max}$ is determined from intercept of line with y-axis $(1/v_o)$

Looking for extra marks?

Enzyme assays are important clinically, e.g. in detecting metabolic disorders by measuring changes in concentration or activity of the enzyme (or isoenzyme) in disease compared with normal state. When tissue enzymes are detected in the blood, this often indicates tissue damage, e.g. following heart attack.

3.3 ENZYME INHIBITION

An enzyme inhibitor (I) is a substance that binds to an enzyme and interferes with its activity, either by preventing formation of an ES complex or preventing conversion of the ES complex to E + P. Inhibition may be reversible or irreversible.

Classification of inhibitors

- **Irreversible inhibitors** become covalently bound to enzyme (Figure 3.5)
    - i.   cannot be removed by dialysis or dilution
    - ii. give time-dependent inhibition
    - iii. enzyme may be protected by substrate (if active site modified)
- **Reversible inhibitors** bind to enzyme by same non-covalent forces that bind substrate and product to enzyme (Figure 3.5)
    - i.   can be removed, e.g. by dialysis or dilution
    - ii. give instantaneous inhibition
    - iii. may be competed out by substrate (if active site occupied)

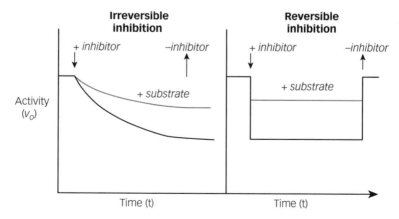

Figure 3.5 Irreversible and reversible inhibition of enzymes.

Irreversible inhibitors

- Inhibitor I forms a stable covalent bond with E
- Usually via alkylation or acylation of side chain of amino acid in active site, e.g.
    i. organophosphate insecticides: inhibit *chymotrypsin* (*serine proteases*)
    ii. nerve gas DFP (di-isopropyl fluorophosphate): inhibits *acetylcholinesterase* (a *serine esterase*)
    iii. penicillin inhibits *glycopeptidetranspeptidase* (blocks bacterial cell wall synthesis)
    iv. mercuric ions bind to sulphydryl groups on proteins (heavy metal toxicity)
- Irreversible inhibitors can be used to identify (mark) active site residues

Reversible inhibitors

- Reversible inhibitors can be divided into two types (Table 3.3):
    i. inhibition is overcome by adding increased substrate concentrations (**competitive inhibition**)
    ii. inhibition is not overcome by adding more substrate

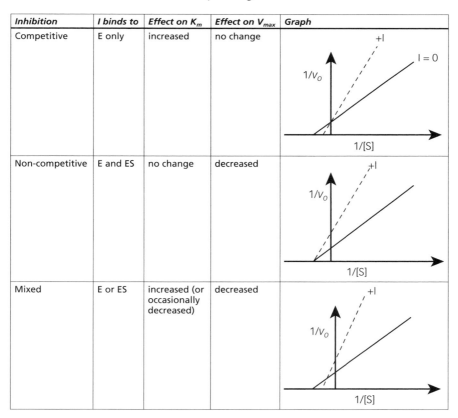

Inhibition	I binds to	Effect on K_m	Effect on V_{max}	Graph
Competitive	E only	increased	no change	
Non-competitive	E and ES	no change	decreased	
Mixed	E or ES	increased (or occasionally decreased)	decreased	

Table 3.3 Kinetic behaviour of reversible enzyme inhibitors

Enzyme inhibition

- In the latter case, only k_{cat} may be affected (non-competitive inhibition), or both k_{cat} and K_m (mixed inhibition)
- K_i = dissociation constant of I from EI complex

Competitive inhibition

- I can bind E only if no other ligands bound to E:
    - i. I binding blocks S binding to E
    - ii. S binding blocks I binding to E
    - iii. S and I compete for binding to E
- Formation of EI complex can be reversed by increasing concentration of S
- Hence, V_{max} is the same ± inhibitor
- I usually resembles S structurally and/or chemically
    - i. benzamidine = competitive inhibitor of *trypsin*—mimics free amine side groups of lysine and arginine in a polypeptide chain (used in biochemical purifications to inhibit proteolysis)
    - ii. ethanol competes with ethylene glycol for *alcohol dehydrogenase* (used to treat ethylene glycol poisoning)
- Classical competitive inhibition: S and I both bind active site of E
- Effects of chemical changes on affinity of enzyme for I can be used to assess the 'shape' of active site
- When S and I both present, proportion of E able to form ES complex depends on:
    - i. relative affinity of E for S and I (relative values of K_m and K_i)
    - ii. concentrations of S and I present ([S], [I])
- K_i can be determined from its effect on the apparent value of K_m:

$$K_m^{app} = K_m(1 + [I]/K_i)$$

Non-competitive inhibition

- I binds to either E or ES complex, forming *inactive* EI or ESI complexes
- Inhibition is NOT relieved with excess S
- I effectively titrates active enzyme from solution
- I is not normally a substrate analogue and binds a different site on E from S (probably allosteric)
- In allosteric non-competitive inhibition, I alters conformation of E so that S can still bind but E can no longer catalyse the reaction
- V_{max} decreases ($1/V_{max}$ increases) with no change in K_m
- K_i can be determined from its effect on the apparent value of V_{max}:

$$V_{max}^{app} = V_{max} / (1 + [I]/K_i)$$

Mixed inhibition

- Generally allosteric
- V_{max} decreases ($1/V_{max}$ increases) as some E forms inactive EI or ESI complexes
- K_m changes as equilibrium is shifted between ES and ESI complex formation
- Inhibition is NOT relieved by excess S

 Check your understanding

Under what conditions is the Michaelis–Menten relationship an appropriate description of enzyme behaviour? Why might it break down *in vivo*? (*Hint: what assumptions are made in deriving this equation? Do they hold in the cell?*)

How can competitive and non-competitive inhibitors be distinguished from one another? (*Hint: Michaelis–Menten kinetics.*)

3.4 ENZYME MECHANISMS

Most enzymes carry out relatively simple chemical transformations. Complex transformations involve a number of simple steps which together constitute a biochemical pathway. A subset of amino acid side chains of the enzyme frequently act as acid/base catalysts or nucleophiles, and/or hydrogen bond to the transition state to stabilize it. Certain cofactors (such as metal ions or vitamin derivatives) confer additional chemical versatility on enzyme active sites.

Catalytic groups in active sites

- Most amino acids in enzymes do not contribute directly to catalysis
- Aliphatic and aromatic amino acids may aid in binding substrate by **hydrophobic interactions**
  - see *1.2 Amino acids (p. 6) for classification of amino acids*
- Lysine and arginine are permanently positively charged and may aid in binding substrate by ionic interactions but cannot transfer H^+ (although lysine can act as an **electrophile** in *aldolase*)
- Amino acids most commonly involved in chemical catalysis are:
  - i. glutamate and aspartate (acid/base catalysts due to carboxylic acid side chain –COOH group)
  - ii. serine and cysteine (**nucleophiles** due to –OH/–SH side chains)
  - iii. histidine (can act as acid, base or nucleophile)
- Amino acids commonly involved in hydrogen bonding at active sites are:
  - i. five amino acids above (i.e. glu, asp, ser, cys, his)
  - ii. amides: glutamine, asparagine
- Enzymes are pH-dependent because they require active site amino acids to be in a particular ionization state (e.g. glu present as –COOH to act as a proton donor)

Enzyme mechanisms

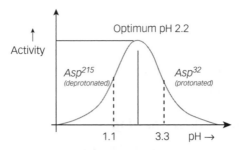

Figure 3.6 pH profile of pepsin activity. For the enzyme to be active, asp[215] must be deprotonated and asp[32] protonated.

- pK$_a$ values of amino acid side chains in enzymes may vary from their pK$_a$ in free solution due to their **microenvironment** (e.g. glu –COOH may not ionize until pH >5 if surrounded by hydrophobic side chains) (Figure 3.6)

Pepsin and the acid proteases

- Catalyse the hydrolysis of peptide bonds in proteins (see Figure 3.7)
- *Acid proteases* have two aspartate residues at the active site
- In human *pepsin*, asp[215] is in the charged (base) form and asp[32] in the uncharged (acid) form
    - i. asp[32]–COOH group donates a proton to the C=O of the peptide bond to be cleaved
    - ii. asp[215]–COO$^-$ group removes a proton from H$_2$O making it more nucleophilic to attack the C=O of the peptide bond
- Acid proteases typically have a pH optimum in the range 2–4
- Examples include *pepsin* (in the mammalian stomach) and *HIV protease* (involved in assembly of the AIDS virus)

Looking for extra marks?

Tight binding competitive inhibitors of the *HIV protease* are used in AIDS therapy (e.g. saquinavir, ritonavir, etc.).

The serine proteases

- Catalyse the hydrolysis of peptide bonds in proteins (see Figure 3.8)
- Utilize a catalytic triad, comprising ser[195]–OH (nucleophile), his[57]–NH (acid), and asp[102]–COO$^-$ (orients the residues) (*chymotrypsin* numbering)
- Mechanism involves a covalently bound intermediate, the acyl enzyme:
    - i. serine–OH is polarized by the his residue and nucleophilically attacks the peptide bond

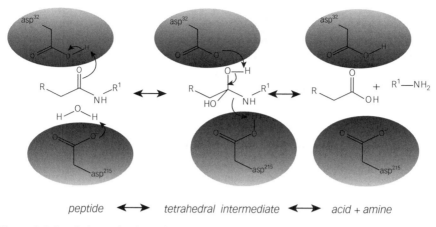

Figure 3.7 Catalytic mechanism of pepsin. R, R¹ refer to remainder of polypeptide chain.

    ii. peptide C-atom attacked changes from trigonal to tetrahedral as the double bond of the carbonyl is broken

   iii. O⁻ produced is stabilized by H-bonding in the 'oxyanion hole'

   iv. his–NH donates a proton to the peptide-N atom, and the amine leaves, resulting in acylation of the serine

    v. hydrolysis of the acyl enzyme occurs via a similar pathway, using H_2O as the nucleophile to displace serine

- Specificity pocket lies adjacent to the catalytic groups. This contains a negatively charged side chain in trypsin, which hence cleaves adjacent to positive side chains (lys, arg)
- Examples include:

    i. digestive proteases (*trypsin, chymotrypsin*)

   ii. blood clotting factors (e.g. *Factor IX, thrombin*)

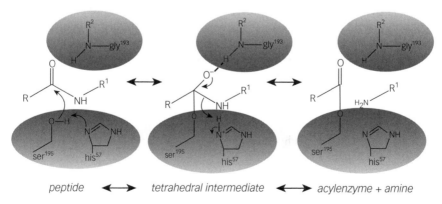

Figure 3.8 Catalytic mechanism of serine proteases. Note hydrogen bond from gly-NH to tetrahedral intermediate—'oxyanion hole'. R¹, R² refer to remainder of polypeptide chain.

Enzyme mechanisms

Triose phosphate isomerase

- Catalyses the isomerization between glyceraldehyde 3-phosphate (GAP) and dihydroxyacetone phosphate (DHAP) in glycolysis (see Figure 3.9)
- Utilizes a glu[165]–COO⁻ side chain as a base (proton acceptor) and a his[95] side chain as an acid (proton donor)
- Mechanism involves a non-covalently bound intermediate, the ene-diol:
  i. glu–COO⁻ abstracts a proton from C2 of GAP and moves it to C1 of DHAP
  ii. his–NH donates a proton to C_1=O of GAP and regains it from C_2–OH of ene-diol
- *Triose phosphate isomerase* catalyses very close to the rate of collision between enzyme and substrate showing a catalytic efficiency of ~100%—a 'perfect' catalyst

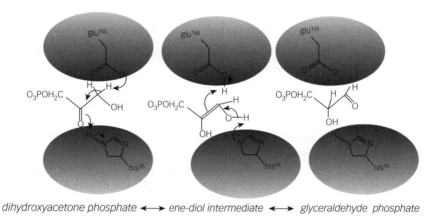

dihydroxyacetone phosphate ⟷ ene-diol intermediate ⟷ glyceraldehyde phosphate

Figure 3.9 Catalytic mechanism of triose phosphate isomerase.

Looking for extra marks?

Triose phosphate isomerase has a characteristic **supersecondary structure** framework—an 'αβ barrel' structure (eight α-helices in a ring surrounding eight parallel β-strands). Active site lies on one end of the barrel. This general arrangement is shared by several other enzymes.

Dehydrogenases

- Catalyses the transfer of a hydride (H⁻) ion from an alcohol to NAD⁺ (or the reverse) (see Figure 3.10)
- Utilize his[195] (*lactate dehydrogenase* numbering) residue as proton acceptor and a hydrophobic patch to polarize NAD⁺
- his[195] is oriented by asp[168]
  i. as nicotinamide binds close to ile[250], positive charge is displaced from ring N atom to opposite end of the ring

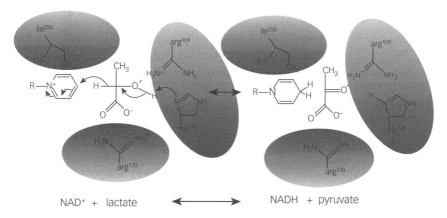

NAD⁺ + lactate ⟷ NADH + pyruvate

Figure 3.10 Catalytic mechanism of lactate dehydrogenase. Note hydrogen bond from arg¹⁰⁹ to carbonyl oxygen which aids polarisation.

   ii. loop (not shown) closes down over reacting species, excluding water

  iii. his removes H⁺ from –OH of the alcohol

  iv. remaining electrons migrate towards C–H grouping, and activated (positive) end of nicotinamide ring picks up H⁻ ion

- Composition of loop determines the specificity: loop in *lactate dehydrogenase* contains neutral residues, but replacement by positive residue converts enzyme to *malate dehydrogenase*
- *Lactate dehydrogenase* has **isoenzymes** that kinetically favour reduction of pyruvate (skeletal muscle isoenzyme) or oxidation of lactate (heart muscle)
- Other examples of dehydrogenases include *alcohol dehydrogenase* and *glyceraldehyde phosphate dehydrogenase*

 Check your understanding

Why are histidine (his) residues often found at active sites? (*Hint: think about the chemical properties of the his side chain and why this might be useful.*)

Why are enzymes normally made out of protein? (*Hint: which amino acid residues contribute useful properties to the catalytic site?*)

3.5 ENZYME REGULATION

The activity of many enzymes can be altered depending on requirements of the cell.

Key concepts in enzyme regulation

- Pathways are controlled by regulating the activity of the enzymes involved
- Only a limited number of enzymes are directly regulated

continued

Enzyme regulation

- Enzymes may be regulated by:
    - i. non-covalent interactions (allosteric effects), or
    - ii. covalent changes (proteolysis, phosphorylation)
    - iii. altering the amount of enzyme present

Regulatory enzymes

- Usually act at **committed step** of metabolic pathway (first unique step in pathway)
- Flux of reactants through subsequent enzymes in pathway is therefore controlled by the rate at which their substrate is supplied
- Conserve energy and substrates by acting at early step in pathway
- Activity controlled by altering amount or activity of enzyme
- Altering <u>amount</u> of E:
    - i. alters V_{max} since $V_{max} = k_{cat}[E]$
    - ii. occurs via increased rate of synthesis of enzyme (increased gene expression) to increase levels of E
    - iii. gives rapid control in prokaryotes (minutes) but is slower in eukaryotes (hours)
- Altering <u>activity</u> of E:
    - i. alters k_{cat} or K_m
    - ii. occurs via binding of metabolites to enzyme in **allosteric** control, or phosphorylation in hormonal control
    - iii. gives very rapid response (seconds)
    - iv. in allosteric regulation, control is by molecular effectors other than [S] or [P] (often product of entire pathway) (e.g. citrate negatively regulating *phosphofructokinase* in glycolysis).

    ➧ *see 5.2 Glycogen breakdown and synthesis (p. 131) for control of glycolysis*
    - v. in regulatory phosphorylation, *kinases/phosphatases* are used to add/remove phosphate groups

    ➧ *see Tables 5.2 (p. 135), 5.3 (p. 137) and 5.7 (p.146) for regulation of glucose and glycogen metabolism*

Allosteric enzymes

- Sensitive to inhibitors and activators
- Often have multiple identical subunits (quaternary structure)
- Regulatory site = **allosteric site** = effector binding site: physically distinct from enzyme active site (often on different domain or even different subunit of enzyme)
- Allosteric modulator (effector) can be activator or inhibitor:

    i. binds to E non-covalently to give conformational change in E which is transmitted to its active site

    ii. not altered chemically by E

    iii. rarely resembles substrate or product

- Many effectors affect K_m, some affect k_{cat} (e.g. activators reduce K_m or increase k_{cat})
- Sensitive to inhibitors and activators (see Table 3.4)

Pathway	Enzyme	Allosterically activated by	Allosterically inhibited by
Glycolysis	phosphofructokinase 1 (PFK1)	AMP, fructose 2,6-bisphosphate	ATP-citrate
Hormone signalling (e.g. adrenaline)	protein kinase A	cAMP	Not applicable

Table 3.4 Key examples of allosteric control

Allosteric enzymes with multiple subunits are cooperative

- Usually have at least one substrate for which $v_o/[S]$ plot is sigmoidal rather than hyperbolic (i.e. not classical Michaelis–Menten)
- Sigmoidal curve due to cooperativity of substrate binding to multiple binding sites (second substrate molecule binds more readily than first)
- Cooperativity increases 'sensitivity' to substrate/effector (large percentage change in effect for small percentage change in effector), e.g. O_2 binding to haemoglobin

Concerted model of allostery/cooperativity (Monod–Wyman–Changeux)

- One binding site for ligand per enzyme subunit
- Each subunit can have two conformations:
  - i. R (relaxed)—high affinity for ligand/high activity
  - ii. T (tense)—low affinity for ligand/low activity
- Enzyme always retains overall symmetry, i.e. if one subunit undergoes conformational change, all do ('all or nothing')—R_n or T_n but never $R_x T_{n-x}$
- Ligand binding to enzyme shifts the equilibrium between R and T states ($R_n \leftrightarrow T_n$), e.g. substrate/activator stabilizes R_n, inhibitor stabilizes T_n form
- Simplistic but provides useful mathematical prediction of properties

Looking for extra marks?

- PFK has four identical subunits and four binding sites for fructose 6-phosphate (F6P)
- binding of F6P is cooperative, i.e. gives sigmoidal (non-Michaelis–Menten) curve
- X-ray crystallography shows R_4 and T_4 as symmetrical structures

Control by covalent modification

- Slightly slower than allosteric control
- Often functionally reversible—but needs one enzyme for activation (converter enzyme) and a different enzyme for inactivation, e.g. *kinase* and *phosphatase* (see Table 3.5)
- Converter enzymes:
    i. usually allosterically controlled, e.g. by second messengers
    ii. can be covalently modified themselves
- Enzymes with interconvertible R (high affinity, high activity) and T (low affinity, low activity) states may be fixed in either R or T by covalent modification, e.g. phosphorylation on serine, threonine or tyrosine residues by a ***protein kinase***, or dephosphorylation by a ***protein phosphatase***
- ATP is the phosphate donor in phosphorylation
- Equilibrium in ATP-dependent phosphorylation is well over to the right, therefore many molecules of enzyme can be converted by one molecule of kinase (amplification of signal)
- Often several kinases act sequentially, giving a **cascade** of amplification, e.g. *glycogen phosphorylase* is activated by phosphorylation, involving *protein kinase A* (cAMP-dependent) followed by *phosphorylase kinase*. It is inactivated by dephosphorylation by a *phosphatase*
    ➲ *see 5.5 Glycogen breakdown and synthesis (p. 145 and Table 5.7) for control of glycogen metabolism*
- Some enzymes (e.g. *trypsinogen*, blood clotting factors) are activated by proteolysis. This cannot be reversed

3.6 MULTIENZYME COMPLEXES

Key concepts of multienzyme complexes

- Allow regulation of series of related reactions— **metabolic channelling**
- Product of one reaction forms substrate of next reaction
- Transfer occurs between active sites without entering bulk solvent
- Increases reaction rates by:
    (i) increasing local concentration of substrate;
    (ii) preventing problems with insoluble intermediates (e.g. in fatty acid metabolism);
    (iii) protecting labile intermediates from degradation (e.g. *carbamoyl phosphate synthetase*)
- Individual enzymes may be non-covalently associated (e.g. enzymes required for DNA replication)
- Several active sites may be present on same polypeptide chain (e.g. *PFK2* and *FBPase2*)

Pathway	Enzyme	Activated by	Converter	Converter activated by	Inactivated by	Converter	Converter activated by
Control of glycolysis	phosphofructokinase 2	deP	phosphatase		P	protein kinase A	cAMP
Control of glycolysis	fructose bisphosphatase 2	P	protein kinase A	cAMP	deP	protein phosphatase	always on
Glycolysis/TCA cycle	pyruvate dehydrogenase (PDH)	deP	PDH phosphatase	Ca^{2+}	P	PDH kinase	several metabolites
Glycogen degradation	glycogen phosphorylase	P	phosphorylase kinase	P	deP	protein phosphatase	always on (constitutive), also induced indirectly by insulin
Glycogen synthesis	glycogen synthase	deP	protein phosphatase 1		P	phosphorylase kinase	P
						protein kinase A	cAMP (inactivated by phosphodiesterase which converts cAMP to AMP)

Table 3.5 Key examples of regulation by enzyme phosphorylation (P) and dephosphorylation (deP)

Fatty acid synthase (mammals)

- **Homodimer** of 2×230 kDa proteins
- Each contains the five domains needed to elongate fatty acids
  - ➔ *see 5.6 Lipids (p. 152) for fatty acid synthesis*
- Functional domains arranged from N-terminus include:
  - i. *β-ketoacyl synthase*
  - ii. *malonyl/acetyl transferase*
  - iii. *β-hydroxyacyl dehydratase*
  - iv. *enoyl reductase*
  - v. *β-ketoacyl reductase*
- A flexible arm (**acyl carrier protein**, ACP) is attached to the *ketoacyl reductase* and bears pantotheine side chain—intermediates in synthesis are linked to this via a sulphydryl group
- ACP carries substrate from one active site to another (via flexible phosphopantotheinyl unit)—permits coordination and reduces time taken for substrate to 'find' active site—hence increases reaction rate
- Two monomers are assembled in an X shape, with the two groups of active sites surrounding a cleft at each side (see Figure 3.11)
- Transesterase releases the product (palmitoyl CoA)
- In bacterial systems, separate enzyme activities are on separate peptides, rather than separate domains, but ACP still carries intermediates around each active site in turn

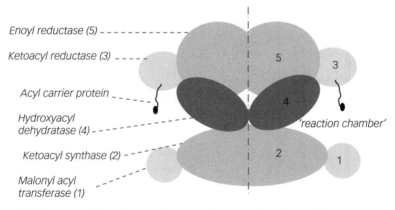

Figure 3.11 Cartoon of domain organization of mammalian *fatty acid synthase* complex. The complex is a dimer of two polypeptides, each containing all the five activities shown. The order of activities is shown (1–5), starting from introduction of a 2C fragment (via malonyl group) to the final reduction of the unsaturated chain prior to repeating the cycle with another 2C fragment. The growing fatty acid chain is bound to acyl carrier protein, which forms a mobile arm bound to the ketoacyl reductase.

Pyruvate dehydrogenase

- Complex of three enzymes (E_1, E_2, and E_3)
- Sequence of reactions is as follows (Figure 3.12):
    i. E_1 bears thiamine pyrophosphate (TPP) **cofactor** and catalyses decarboxylation of pyruvate forming hydroxyethyl TPP ('active acetaldehyde')
    ii. E_1 transfers active acetaldehyde to E_2 where it is bound to a (covalently attached) lipoic acid cofactor via sulphydryl group
    iii. electrons are transferred to reduce disulphide form of lipoic acid to its thiol form, and substrate is oxidized to acetyl group
    iv. E_2 transfers acetyl group onto sulphydryl group of coenzyme A. Acetyl CoA is released
    v. reduced lipoic acid (dithiol form) is oxidized back to disulphide by E_3, which bears *FAD cofactor*
    vi. E_3 oxidizes $FADH_2$ using *NAD^+* as a coenzyme, and NADH is released

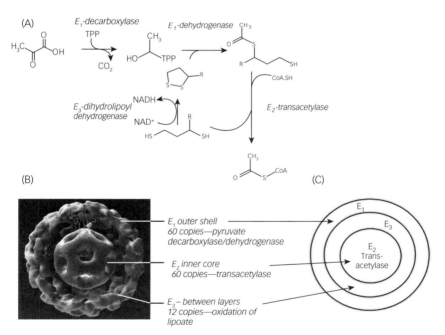

Figure 3.12 Pyruvate dehydrogenase. (A) Reactions carried out by the enzyme complex. (B) Structural arrangement of proteins in pyruvate dehydrogenase complex. (C) Schematic diagram to show this arrangement quickly for exams.
(B) From Milne et al. (2002) *Molecular architecture and mechanism of an icosahedral pyruvate dehydrogenase complex: a multifunctional catalytic machine*, EMBO J 21: 5587–98. Reproduced with permission of Nature Publishing Group.

Multienzyme complexes

- E_2 bears lipoic acid side chain and acetyl intermediate is linked to this via a sulphydryl group
- Lipoic acid carries substrate from one active site to another (via a flexible link)
- Mammalian *pyruvate dehydrogenase* has a core of 60 copies of E_2 arranged in an icosahedron, surrounded by a shell of (up to) 60 E_1 enzymes. Copies of E_3 are located between the two
- Phosphorylation of E_1 inactivates enzyme and is used in its regulation in mitochondria

 Check your understanding

Short answer questions

What are the advantages of multi-enzyme complexes? (*Hint: think about concentrations of reactants, possibilities of side reactions and control.*)

Which steps in a pathway are commonly subject to regulation? Give a specific example. (*Hint: look for irreversible reactions that often involve an input of energy.*)

Longer essay questions

What are the key features of enzyme-mediated catalysis?

How can enzyme activity be regulated? (*Hint: remember gene regulation (changes [E]) as well as allosteric and covalent modification.*)

Give an example of a transition state analogue for an enzyme. Why are transition state analogues good inhibitors of enzymes? (*Hint: how do enzymes stabilize transition states?*)

4 Genome stability and gene expression

Key concepts

- Genetic information is stored in DNA in the order of nitrogenous bases
- Information in DNA must be accurately copied ready to be passed to the next cell generation—**DNA replication**
- To access the information encoded in DNA, a relatively short-lived copy of the genetic information is made by **transcription** into RNA
- The three-letter nucleotide code of messenger RNA is translated into the amino acids of protein
- **Translation** takes place in the ribosome using adapter tRNAs that recognize both the three-letter nucleotide code and amino acids specified by that code
- Central dogma of molecular biology is therefore DNA→ RNA→ protein (Figure 4.1)
- Exceptions to the central dogma include **reverse transcription** of RNA into DNA by retroviruses
- Any errors in the bases of DNA must be corrected by specific **DNA repair** processes for the integrity of the genetic code to be maintained
- DNA regions can be cut and rejoined differently to alter genome sequence, e.g. in sexual reproduction, by **DNA recombination**

DNA replication

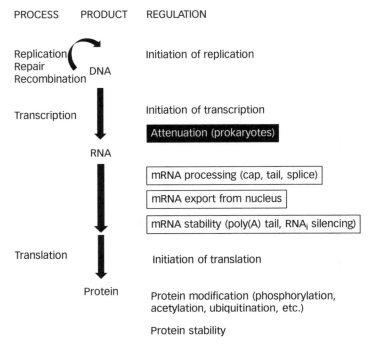

Figure 4.1 Levels of regulation of genome stability and gene expression. Core biochemical processes are shown on the left, with stages of regulation on the right. Processes specific to eukaryotes are outlined in black; those found only in prokaryotes are in black box.

4.1 DNA REPLICATION

DNA replication is the accurate and precise copying of DNA template by DNA polymerase.

Key features of DNA replication

- One double-stranded DNA (dsDNA) molecule is used to generate two identical double-stranded molecules
- Template-directed process using information present in template strand of DNA to synthesize a complementary strand
- Synthesis of new strand of DNA is catalysed by *DNA polymerase*
- **Phosphodiester bonds** are formed on addition of dNTP substrates
- Energy is released by pyrophosphate hydrolysis
- In eukaryotes, it usually occurs once and once only per cell division cycle

Aspects of DNA replication

- Replication starts at defined sites in the double-stranded DNA template called **origins of replication**
- Replication is:
    i. **Semiconservative**, i.e. new duplex DNA contains one strand of parental DNA and one daughter strand
    ii. Semidiscontinuous, i.e. **leading strand** synthesized continuously, while **lagging strand** is made as short regions that are then joined by *DNA ligase*
    iii. **Bidirectional** from origin (see Figure 4.2)
- Both strands of duplex DNA are copied at same time, forming a **replication fork:**
    i. each parental strand acts as a template for synthesis of a complementary strand of daughter DNA, defined by base pairing rules (A = T, G = C)
    ii. hence mechanism of DNA replication is closely linked to DNA structure
- Mediated by *DNA-dependent DNA polymerases* which synthesize DNA in 5′ to 3′ direction with respect to new (nascent) strand

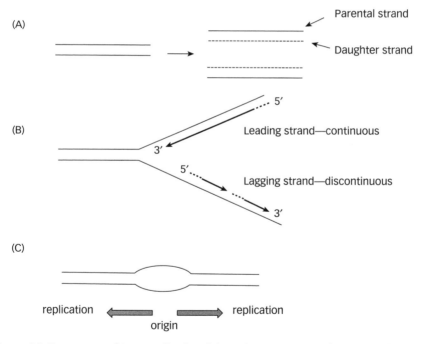

Figure 4.2 Key aspects of DNA replication. (A) Semiconservative replication: each new duplex contains one intact parental strand of DNA and one new complementary daughter strand. (B) Semidiscontinuous replication: the leading strand is made continuously in the 5′–3′ direction and the lagging strand is synthesized as short Okazaki fragments (again 5′–3′) that are subsequently joined together. (C) Replication is bidirectional, i.e. proceeds in both directions away from an origin (some exceptions, e.g. certain viruses).

DNA replication

- Three separate stages: (i) initiation; (ii) elongation; (iii) termination
- Very high fidelity—error rate approximately 1 in 10^9 nt—due to tight enzyme active site and additional proof-reading activity of *DNA polymerases*
- Replication is *regulated* to occur once and only once per cell division cycle (except for viruses, some plasmids, and organelle genomes)
- Entire genome copied—not just genes
- The major steps in replication are essentially the same in prokaryotes and eukaryotes (see Table 4.1)
- Exceptions to basic rules in some **bacteriophage** (e.g. M13), eukaryotic viruses (e.g. polyomavirus) and genomes of chloroplasts and mitochondria

Process	Prokaryotic enzyme	Eukaryotic enzyme
Initial DNA synthesis from RNA primer	DNA pol III	DNA pol α
Leading strand synthesis	DNA pol III	DNA pol ε
Lagging strand synthesis	DNA pol III	DNA pol δ
Okazaki fragment processing	DNA pol I and III	DNA pol δ
Mitochondrial DNA replication	n/a	DNA pol γ

Table 4.1 Functions of replicative DNA polymerases

DNA polymerases

- Also known as *DNA pol* or *DNAP*
- DNA template-dependent enzymes: incoming nucleotide only fits enzyme's active site if complementary to nucleotide on template strand
- Do not synthesize DNA *de novo*—require RNA primer
- Catalyse the formation of phosphodiester bonds (Figure 4.3) via nucleophilic attack by lone pair electrons of 3' OH from primer (or newly incorporated nucleotide) onto α phosphate of incoming **deoxyribonucleotide triphosphate** (dATP, dTTP, dCTP, and dGTP)
- ➔ *see 1.5 Chemical composition of DNA (p. 20)*
- Overall reaction: $DNA_n + dNTP \rightarrow DNA_{n+1} + PP_i$
- Large negative ΔG (≈ -40 kJ/mol) as reaction driven forwards by hydrolysis of PP_i to $2P_i$ by *pyrophosphatase* (i.e. thermodynamically favourable)
- All known *DNA pols* polymerize DNA in a 5'→3' direction with respect to nascent strand
- *DNA pol I* and *III* in *E. coli* and DNA polymerases δ, ε, and γ in eukaryotes have 3'→5' *exonuclease* activity, i.e. 'proof-reading' to remove incorrectly added bases
- Often multi-subunit complexes e.g. polymerase, 5'→3' exonuclease (repair function), and 3'→5' exonuclease (proof-reading function) activities can be present on different subunits of same enzyme

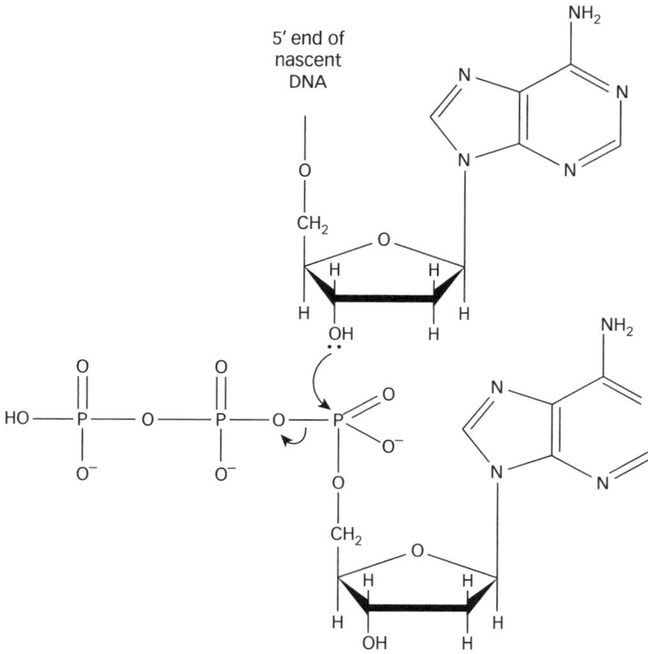

Figure 4.3 Mechanism of phosphodiester bond formation. Note that the pyrophosphate released is very quickly hydrolysed to inorganic phosphate, a favourable reaction with very negative ΔG (Gibbs free energy). Hence no energy input is required for phosphodiester bond formation (though synthesis of dNTP precursors requires energy— see Chapter 7).

- Polymerization rates:
     i. ~500 nt/second in prokaryotes
     ii. ~50 nt/second in eukaryotes
- Leading strand polymerases have high **processivity**, i.e. many nucleotides polymerized each time the polymerase associates with DNA template
- Processivity can be enhanced by binding to a sliding clamp that holds the *DNA pol* on the DNA

Steps in DNA replication

1. Initiation
- Replication **origin** is a specific sequence in DNA, e.g. OriC in *E. coli*, ARS in *S. cerevisiae*, or defined by a combination of sequence, chromatin structure, and possibly epigenetic modification in higher eukaryotes.
- Single origin in *E. coli*; multiple origins in eukaryotes
- Origin sequences are usually very AT-rich (see Figure 4.4)

DNA replication

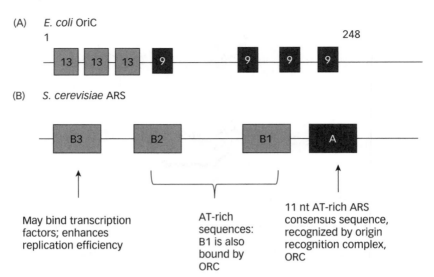

(A) *E. coli* OriC

(B) *S. cerevisiae* ARS

May bind transcription factors; enhances replication efficiency

AT-rich sequences: B1 is also bound by ORC

11 nt AT-rich ARS consensus sequence, recognized by origin recognition complex, ORC

Figure 4.4 Replication origins. Sequence-specific origins comprise AT-rich 'boxes' important in binding to origin recognition proteins. (A) In *E. coli*, the single origin, OriC, is made up of 3 x 13mer boxes and 4 x 9mer boxes of similar but not identical sequence: the origin DNA can bend round a multimer of DnaA protein to open up the duplex. The minimal core origin is 248 nucleotides long. (B) In the budding yeast *S. cerevisiae*, the origin (autonomously replicating sequence, or ARS) is also AT-rich and modular, with four conserved and essential boxes, two of which are directly involved in binding to the origin recognition complex, ORC.

- Origin recognition factor(s) (DnaA in *E. coli*, ORC in eukaryotes) bind to replication origin (DNA)
- Duplex DNA opens via torsional strain upon protein binding (AT-rich so only two H-bonds between each base pair—easier to open the duplex)
- *Helicase* is recruited by helicase loaders—unwinds duplex further using energy from ATP hydrolysis to break H-bonds between base pairs in template DNA
- ssDNA is stabilized by single-stranded DNA (ssDNA) binding protein
- *Primase* is recruited which synthesizes a short RNA primer (~7–10 nt)
- In eukaryotes, *primase* is tightly associated with *DNA pol α* which synthesizes a short stretch (~20 nt) of initiator DNA (iDNA)
- Sliding clamp protein is loaded by clamp loader at the primer–template junction (i.e. at the end of the primer/iDNA)
- Replicative *polymerases* are recruited by binding to sliding clamp → elongation

2. Elongation
- Incoming dNTP base-pairs with template DNA within active site of *DNAP*
- Replicative *DNA polymerases* catalyse phosphodiester bond formation between 3' OH of previous dNMP (or NMP of RNA primer) and 5' phosphate of incoming dNTP (nucleophilic attack) (see Figure 4.3)

- Further ATP-dependent unwinding of duplex DNA by *helicase* ahead of DNA polymerases opens up a replication fork (see Figure 4.5)
- Torsional strain from unwinding relieved by *topoisomerases* (positive supercoils) and *gyrases* (negative supercoils)
- Leading strand DNA synthesized continuously in 5′–3′ direction
- On lagging strand, DNA is synthesized 5′–3′ as short Okazaki fragments (~100–200 nt eukaryotes, ~1,000–2,000 nt prokaryotes). *DNA pol* dissociates from the end of a completed fragment and then binds at 3′ end of new RNA primer closer to the replication fork to synthesize the next Okazaki fragment
- RNA primers removed, possibly also with removal of some DNA (especially iDNA synthesized by *DNA pol* α in eukaryotes)
- Gap left on removal of RNA primer is filled with DNA by *DNA pol* (see Table 4.1)
- Nick in sugar-phosphate backbone sealed by ***DNA ligase***

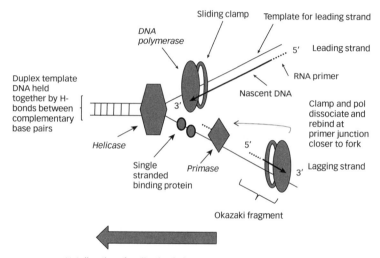

Figure 4.5 The replication fork. Duplex DNA is opened up by *helicase* using the energy from ATP hydrolysis to break H-bonds between base pairs. Both strands serve as templates for synthesis of complementary DNA by *DNA polymerase* in association with a sliding clamp protein. Single-stranded DNA on the lagging strand is stabilized by binding to SSB. *Primase* synthesizes RNA primers complementary to the DNA sequence, then *DNA polymerase* takes over and synthesizes DNA, in long stretches on the leading strand, and as shorter Okazaki fragments on the lagging strand, which are then processed to remove the RNA and joined by *DNA ligase*.

3. Termination

- *Topoisomerases* resolve concatemers or intertwined forks
- In *E. coli*, Tus protein bound to Ter site acts as 'wedge' to slow down fork movement in one direction, allowing resolution of replication intermediates by *topoisomerase* and *gyrase*

- In eukaryotes, possibly similar termination sites where protein is bound to replication fork barrier
- Ends of linear eukaryotic chromosomes are replicated by action of *telomerase* which adds on short repetitive DNA sequences synthesized according to the enzyme's internal RNA template, to form **telomeres**

Accuracy of DNA replication

- Accuracy is achieved by base pairing between the template strand and dNTPs, and by helical geometry
- *DNA pol* active site excludes solvent (high enthalpy, low entropy)—increases energy differences between correct and incorrect base, favouring accuracy
- If the wrong base is added, proof-reading activity of *DNA pols (3′→5′ exonuclease* activity) will usually remove the incorrect base (DNA moves 40° between polymerization active site and *exonuclease* active site)
- If there is still a mistake, DNA repair machinery will correct a mismatch, by mismatch repair (MMR)

  ⮕ *see 4.4 DNA repair (p. 112) for more details on DNA repair processes.*

  In this case it is important to correct the newly synthesized strand, or a mutation will be introduced. In bacteria the 'old' strand is marked by methylation

Regulation of DNA replication

- Three key mechanisms in prokaryotes such as *E. coli* act to regulate replication origin firing:
  i. RIDA: regulatory inactivation of DnaA, i.e. conversion of DnaA from active ATP-bound form to inactive ADP-bound form by protein Hda
  ii. SeqA: sequesters the newly replicated origin (hemimethylated) so it cannot act as a replication start site
  iii. datA: a DNA sequence with similarity to the boxes in OriC, that binds and titrates DnaA protein
- In eukaryotes, replication origin firing is regulated at two main stages:
  i. licensing (i.e. assembly of the *helicase* onto the origin) predominantly by regulating activity of ORC and levels of helicase loaders Cdc6 and Cdt1
  ii. origin firing (i.e. recruiting the replicative *polymerases* and starting the process of DNA synthesis) via phosphorylation by *cyclin-dependent kinases (Cdks)*

Evidence for DNA replication

- Semiconservative replication shown by CsCl density gradient centrifugation of ^{15}N-DNA replicated in media with ^{14}N (Meselsohn and Stahl): newly synthesized DNA has one heavy (^{15}N) and one light (^{14}N) strand (Figure 4.6)

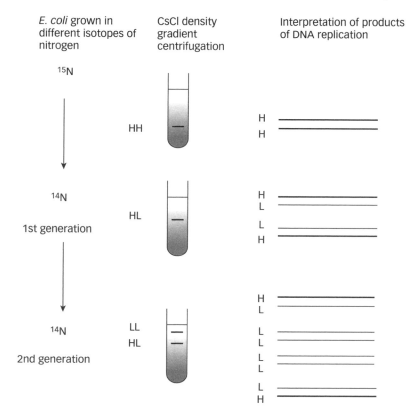

E. coli grown in different isotopes of nitrogen

CsCl density gradient centrifugation

Interpretation of products of DNA replication

Figure 4.6 Replication is semiconservative. *E. coli* are grown in 'heavy' nitrogen (^{15}N) so all genomic DNA has ^{15}N in both strands of the duplex. DNA is extracted and centrifuged through a gradient of very dense caesium chloride to separate it according to density. Parental DNA runs heavy–heavy (HH), i.e. ^{15}N in both strands. After one round of DNA replication in 'light' (^{14}N) nitrogen, each DNA duplex is made up of one parental heavy strand and one nascent light strand, and sediments at a heavy–light (HL) position. After a second round of replication, half the DNA is light–light (i.e. has ^{14}N in both strands, since the 'nascent' DNA from round one is now a 'parental' template for synthesis in round 2). The other two molecules are HL with ^{15}N in one of the strands inherited from the original parental DNA (Meselson and Stahl).

- Semidiscontinuous replication shown by ^{3}H-T pulse-labelling replicating DNA, then separating by alkaline sucrose density gradient centrifugation (Okazaki) (Figure 4.7)
- Bidirectional replication shown by pulse-labelling replicating DNA with high and low activity ^{3}H-thymidine (^{3}H-T), then DNA fibre spreads examined by electron microscopy (EM) with and without silver emulsion (shows radioactivity) (Figure 4.8)
- Bacterial and yeast genetics show which protein factors are necessary, e.g. *DNA pol III* mutants in *E. coli* cannot replicate DNA
- *DNase 1* footprinting shows that replication origins are bound by proteins

DNA replication

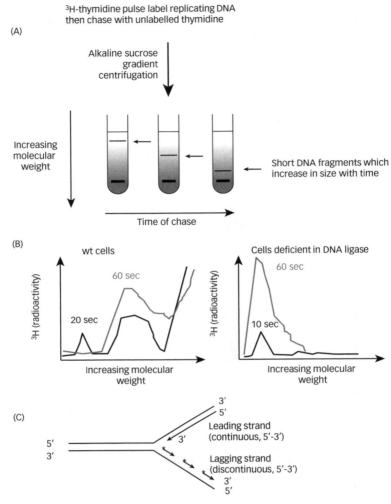

Figure 4.7 Replication is semidiscontinuous. (A) Immediately after a short pulse of
³H-thymidine, which is incorporated into nascent DNA (on leading and lagging strands),
the DNA double helix is disrupted using alkali and the fragments separated on a sucrose
gradient by centrifugation. (B) Very small radioactive fragments were seen at early
times, which shifted to higher molecular weight DNA with time. This change in
molecular weight did not happen in cells lacking *DNA ligase*. (C) Model of the replication fork derived by Okazaki on the basis of this experiment: the lagging strand is made
as short discontinuous Okazaki fragments that are then later processed to remove the
RNA primer (asterisk) and joined by *DNA ligase* to make full length DNA.

- X-ray crystallography shows helical distortion of origin DNA when protein
  bound (e.g. archaeal ORC on ARS)
- X-ray crystallography of proofreading *DNA polymerases* shows tight fit in active
  site and separate active site for *exonuclease*

(A) Electron microscopy

(B) Fibre autoradiography

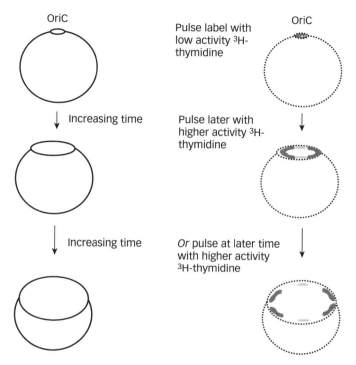

Figure 4.8 The *E. coli* genome replicates bidirectionally from a single origin. (A) In electron microscopy (EM), the DNA forms a theta (θ) structure consistent with replication from one origin (OriC). (B) Replicating cells are pulse labelled with radioactive nucleotides (³H-thymidine) first of low specific activity then with higher specific activity. Then the DNA is isolated, coated with photographic emulsion, and visualized by EM to show the site of radioactive ³H-T incorporation. The finding that high activity ³H-T flanks the low activity label on both sides of the replication origin shows that DNA synthesis happens at *both* forks in the theta structure, so the replication goes in both directions at once, i.e. it is bidirectional. (Note that the DNA is not visible with this technique so is shown as a dotted line.)

- DNA replication can be reconstituted *in vitro* with purified components
- Inhibitors of DNA replication can be used to probe critical steps:
  - i. ATPγS blocks *helicase*-dependent unwinding of duplex DNA
  - ii. nalidixic acid inhibits prokaryotic *DNA gyrase* (prokaryotes)
  - iii. aphidicolin inhibits eukaryotic replicative *polymerases* (dCTP mimic)
  - iv. hydroxyurea depletes nucleotide pools and activates checkpoints (eukaryotes)
- Table 4.2 compares DNA replication in prokaryotes and eukaryotes

Step	Process	Prokaryotic, e.g. E. coli	Eukaryotic, e.g. S. cerevisiae
Initiation	origin of replication	OriC	ARS
	origin recognition	DnaA	ORC1-6
	helicase loader	DnaC	CDC6/cdt1
	helicase	DnaB	MCM2-7
	stabilization of ssDNA	SSB	RPA (SSB)
	RNA primer synthesis	DnaG (primase)	primase
	initial primer extension	DNA pol III	DNA pol α
	clamp loader	γ complex	RFC
	sliding clamp	β subunit (dimer)	PCNA (trimer)
Elongation	leading strand synthesis	DNA pol III	DNA pol ε
	lagging strand synthesis	DNA pol III	DNA pol δ
	co-ordination between strands	τ protein	GINS/cdc45
Okazaki fragment processing (OFP)	primer removal	RNaseH/DNA pol I	RNase H1/FEN1
	gap filling	DNA pol I	DNA pol δ
	nick sealing	DNA ligase	DNA ligase 1
Termination	termination site	Ter	replication fork barrier
	termination protein	Tus	RTF1, fob1
	relief of torsional strain	topoisomerase/gyrase	topoisomerase I and II

Table 4.2 Comparison of prokaryotic and eukaryotic DNA replication

 Check your understanding

How is fidelity in DNA replication achieved? *(Hint: remember base-pairing, enzyme active site, proof-reading, and repair of mismatches after replication has finished.)*

How can both strands of DNA be copied by polymerases which only synthesize DNA in the 5′ to 3′ direction? *(Hint: leading and lagging strands—make sure you draw the replication fork with labelled proteins and show direction of DNA synthesis, and state evidence. Discuss why always 5′–3′, i.e. need release to PP_i and its hydrolysis to $2P_i$ to drive reaction forwards.)*

4.2 RNA SYNTHESIS (TRANSCRIPTION)

Transcription is the DNA-directed synthesis of a complementary single-stranded RNA molecule by *RNA polymerase* using NTP substrates. Products of transcription include rRNA, mRNA, tRNA, and small functional RNAs including miRNA.

Key concepts in transcription

- Transcription starts at specific sequences in the DNA template at or near to **promoters**
- Promoters are short sequences in DNA that bind *RNA polymerase* (directly or through association with **transcription factors**) and allow initiation; they are usually found upstream of the transcription start site

continued

- DNA opens up on protein binding to form **transcription bubble**
- Only one strand of DNA is copied into functional RNA (template strand)
- RNA synthesis is mediated by DNA-dependent *RNA polymerases*
- Nascent RNA is synthesized in 5′ to 3′ direction
- Only defined regions of the genome are copied, i.e. genes
- Three major stages: initiation, elongation, and termination
- Regulated according to cellular need for functional RNAs and proteins: some genes are transcribed at a low level all the time ('housekeeping genes') while others are transcribed only under specific circumstances (e.g. heat shock genes)
- Regulation can be positive or negative (gene actively turned 'on' or actively turned 'off')
- Transcription provides the key regulatory step in gene expression (much more tightly regulated than translation)
- Transcription can take place at any stage of cell cycle except during division (mitosis and meiosis)
- Fundamental mechanisms of transcription are very similar in prokaryotes and eukaryotes but location differs: within cytoplasm of prokaryotes, within nucleus of eukaryotes
- Genes of mitochondrial and chloroplast genomes are also transcribed by similar mechanism
- Eukaryotic transcripts are often processed during or after transcription, e.g. mRNA acquires a 5′ cap, introns are spliced out and a polyA tail added at the 3′ end

RNA polymerases

- Also known as *RNA pol* or *RNAP*
- Synthesize RNA from ribonucleotide triphosphate (NTP) precursors
  - ➜ *see 1.6 RNA (p. 26) for RNA structure*
- Are DNA template-directed and use complementary base pairing to copy one strand of DNA into RNA
- Polymerize nascent RNA in a 5′ to 3′ direction
- Overall reaction: $RNA_n + NTP \rightarrow RNA_{n+1} + PP_i$
- Reaction driven forwards by hydrolysis of PP_i to $2P_i$ (mediated by *pyrophosphatase*)
- Bacterial *RNA polymerase* has a core enzyme important for RNA synthesis; the holoenzyme has an additional σ subunit involved in promoter recognition
- The core contains two copies of the α subunit plus β, β′ and ω subunits (i.e. $\alpha_2\beta\beta'\omega$)
- Role of the subunits:
  - α (alpha)—initiates *RNAP* enzyme assembly (dimer in holoenzyme)

RNA synthesis (transcription)

- ○ β and β′ (beta and beta prime)—catalytic centre
- ○ ω (omega)—*RNAP* enzyme assembly
- ○ σ (sigma)—only present in the holoenzyme and confers promoter specificity
- Eukaryotes have three different types of *RNA polymerase*, each made up of multiple subunits:
    - i. *RNA pol I* → rRNA
    - ii. *RNA pol II* → mRNA (and some snRNAs)
    - iii. *RNA pol III* → tRNAs, 5S rRNA and snRNAs
- Can be regulated to control transcription
- Error rate ~1 in 10^5 nt
- Transcription rate ~50 nt/second (prokaryotes)

Typical prokaryotic gene structure

- Genes encoding related enzymes in same metabolic pathway are often adjacent in the genome and under the control of the same promoter/operator region, i.e. **operon** (see Figure 4.9)
- Transcripts from such gene clusters are **polycistronic** (1 gene = 1 cistron)
- General promoter elements are upstream of genes: these are regions that recruit *RNAP* or factors that assist *RNAP* binding and so act to positively regulate transcription
- **Consensus sequence** of promoter derived by comparison of upstream regions from many different genes:
    - i. −10 (minus ten) box (TATAAT) defines transcription start site at + 1 on DNA and allows duplex melting
    - ii. −35 (minus thirty-five) box is required for efficient transcription
- Other positive regulatory sites exist upstream of some genes, e.g. −65 box for Crp (CAP) protein binding in **lac operon**

  ➔ *see Regulation of transcription in prokaryotes (p. 94)*
- Genes subject to negative regulation have upstream **operator** as well as promoter regions
- The transcription start site (+1) is <u>upstream</u> of the translation start site (AUG start codon): transcript from gene includes 5′ and 3′ untranslated regions (UTR)
- Within 5′ UTR of mRNA transcript is **Shine–Dalgarno sequence**:
    - i. approximately ten bases upstream of translation start codon
    - ii. necessary to define each **open reading frame** in polycistronic message
    - iii. = ribosome binding site or rbs, also known as ribosome entry site or RES
    - iv. purine-rich sequence of 3–9 bases complementary to 16S rRNA of ribosome

  ➔ *see Steps in prokaryotic translation (p. 106) for more information on how the Shine–Dalgarno sequence interacts with the ribosome in protein synthesis*

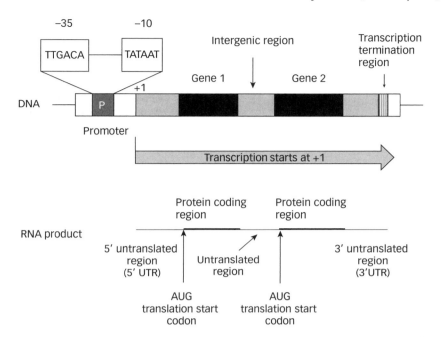

Figure 4.9 Typical structure of a prokaryotic operon. The promoter region consists of two short DNA sequences 10 and 35 bases upstream of the start site of transcription (+1). Binding of *RNA polymerase* to the −10 box places its active site directly over the +1 transcription start site. Transcription begins and continues until a transcription termination region is reached. The mRNA produced contains protein coding regions (or open reading frames) preceded and followed by extra RNA sequences that are not translated into protein, known as 5′ and 3′ untranslated regions (UTRs). There are also untranslated regions between protein coding sequences. (Note that it's OK to draw functional regions of DNA as boxes and regions that are not of interest as lines—trying to represent the double helix here would make the diagram much too complicated.)

Revision tip

Remember that promoter → positive control, operator → genes 'off'

Steps in prokaryotic transcription

1. Initiation

- Promoter DNA is recognized by *RNA pol* σ sigma subunit
- Scanning model: *RNAP* holoenzyme scans ds DNA, forming transient H-bonds (low affinity binding)
- Affinity of σ for duplex promoter is ~10^4 × greater than for non-specific DNA, so associates when it locates the promoter
- When RNAP holoenzyme has associated with promoter element, its binding leads to duplex melting at TATA box (torsional strain)
- σ has lower affinity for ssDNA than dsDNA and dissociates from unwound region

RNA synthesis (transcription)

- Core enzyme begins polymerization of NTPs complementary to base sequence of DNA template strand
- Abortive initiations may occur where a few nucleotides are polymerized, then *RNA pol* stops, and restarts

2. Elongation

- Incoming rNTP forms base pair with template DNA within *RNAP* active site
- *RNA polymerase* core enzyme catalyses phosphodiester bond formation between rNTPs according to base sequence of DNA template strand in 5′ to 3′ direction
- *RNAP* moves along in inchworm fashion
- DNA duplex opens ahead of *RNA pol* and reforms behind it—this is the transcription bubble (see Figure 4.10)

3. Termination

- Two major types: ρ (rho)-dependent and ρ-independent
- ρ-dependent:
    - i. hexameric ρ protein is an *RNA-dependent ATPase*
    - ii. ρ binds onto free 5′ end of transcript
    - iii. ρ wraps RNA around itself to move up the transcript 5′ to 3′
    - iv. when *RNA pol* is paused (e.g. hairpin loop in transcript), *ρ helicase* may break H-bonds between RNA–DNA duplex
    - v. Nus A protein may assist ρ in dissociating *RNA pol* from transcript
    - vi. RNA transcript is released
- ρ-independent:
    - i. depends on RNA sequence (and therefore DNA template sequence)
    - ii. palindromic GC-rich region of transcript forms hairpin loop
    - iii. adjacent run of polyU residues only weakly bonded to template are pulled away on hairpin loop formation
    - iv. transcript and *RNA pol* are released from DNA

Regulation of transcription in prokaryotes

- By σ (sigma) factor: different types of σ subunit of *RNA pol* recognize different types of gene promoters, e.g.
    - i. σ^{70} for consensus −35 and −10 box recognition (many genes)
    - ii. σ^{54} for nitrogen metabolism genes
    - iii. σ^{32} for heat shock genes

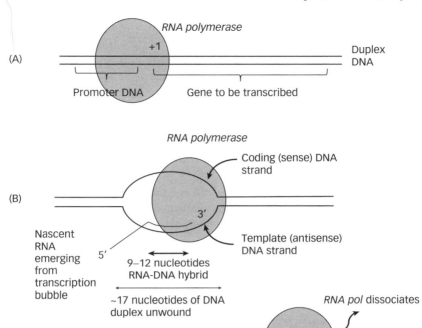

Figure 4.10 Transcription in prokaryotes. (A) Initiation: *RNA polymerase* binds to promoter, which places the enzyme's active site over the transcription start site in the DNA template. (B) Elongation: the DNA duplex is unwound locally, forming a transcription bubble. *RNA polymerase* copies the template (antisense) strand to generate a new (nascent) RNA molecule. (C) Termination: formation of a hairpin loop through G–C base pairs within the nascent RNA destabilizes the RNA–DNA hybrid (also further destabilized by a run of Us in the RNA), leading to dissociation of both the RNA transcript and *RNA polymerase* from the template. NB This shows rho-independent termination. (For clarity, base pairs are not shown.)

- **Inducible** promoter is turned on by the presence of an inducer, e.g. the lac operon (allolactose is inducer)
- **Repressible** promoter is switched off in the presence of a co-repressor, e.g. trp operon (tryptophan = co-repressor)
- **Attenuation** provides fine-tuned negative control, e.g. trp operon
    i. active translation from nascent transcript leads to termination of transcription
    ii. only possible in prokaryotes where transcription and translation take place in same compartment

RNA synthesis (transcription)

1. Transcriptional regulation: the lac operon

- Lac operon = 3 genes encoding proteins for lactose metabolism under control of single promoter/operator
- Glucose is preferred carbon source for *E. coli* so lac operon switched off when glucose is present (see Figure 4.11)
- **Lac repressor** protein (encoded by lac I gene) binds operator (O) upstream of genes encoding lac Z, Y, A, and inhibits transcription—may sterically block *RNAP* binding to promoter (or blocks productive initiation), i.e. <u>negative regulation</u>
- Very low levels of β-gal gene transcription lead to low levels of active *β-galactosidase* in the cell
- In presence of lactose, *β-gal* produces allolactose (isomer of lactose): allolactose is a co-inducer which binds to lac repressor protein and allosterically alters the repressor shape, preventing repressor binding to O. Transcription of the operon is permitted but only occurs at low levels
- When no glucose present, cellular ATP:AMP ratios alter and cAMP (starvation signal) is produced
- cAMP binds to Crp protein which can now bind to −65 box upstream of lac Z, Y, A operon to actively recruit *RNAP* to the promoter and stimulate transcription (Crp = <u>c</u>AMP <u>r</u>egulated <u>p</u>rotein, also known as CAP = catabolite-activated protein/cAMP-activated protein), i.e. <u>positive control</u> of gene expression

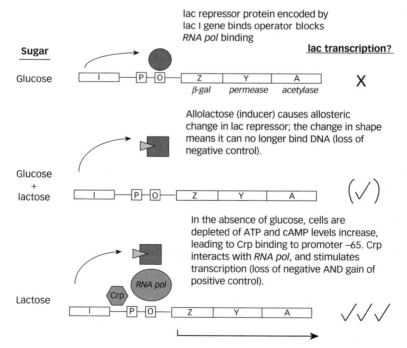

Figure 4.11 The lac operon is under dual negative and positive regulation. P = promoter, O = operator, I = lac I gene encoding lac repressor protein.

- Result: lac operon highly transcribed only when lactose is present (to inhibit repressor) and glucose is absent (to activate CRP)

2. Transcriptional regulation: the trp operon

- Five functional genes within trp operon encode enzymes for tryptophan biosynthesis—energetically costly to transcribe and translate these genes unless tryptophan levels are low (see Figure 4.12)
- Trp repressor protein bound to tryptophan (co-repressor) binds the operator upstream of the trp operon to suppress gene expression (<u>negative control</u>)
- When trp levels fall, trp repressor dissociates, *RNAP* binds promoter, and transcription occurs
- Upstream of the five functional genes but downstream of the promoter lies the trp leader sequence encoding two adjacent tryptophans
- The nascent transcript is translated by a ribosome closely following *RNAP* (since no compartmentalization between transcription and translation in prokaryotes)
- Two possible outcomes:
  - i. if no tRNA^{trp} are present (i.e. low levels of trp amino acid) when the ribosome encounters the leader sequence, it stalls and *RNAP* continues transcribing the functional genes
  - ii. if there is sufficient trp amino acid in cell (hence charged tRNA^{trp}), then the ribosome can synthesize through the leader sequence, leading to formation of a hairpin loop in the RNA that pulls it away from the template DNA. *RNAP* dissociates, i.e. transcription stops (**attenuation**)
    - ➔ *see 4.3 Protein synthesis (translation) (pp. 103–108) for more details on the roles of tRNA and ribosomes in protein synthesis*

Evidence for transcription

- Action of prokaryotic *RNA polymerase* subunits elucidated using specific inhibitors, e.g. rifamycin/rifampicin acts on β subunit of prokaryotic *RNA pol* and blocks formation of first phosphodiester bond—therefore β catalyses bond formation
- *DNase 1* footprinting shows proteins bound to specific sites on DNA, e.g. promoter
- Lac operon:
  - i.   gratuitous inducers (e.g. IPTG) lead to expression of lac Z gene
  - ii. assayed by colorimetric measurement of lac Z activity, e.g. breakdown of artificial chromogenic substrate Xgal produces a blue product
  - iii. lac Z, Y, A genes constitutively expressed on mutation of operator (O) or repressor (I) such that repressor protein cannot bind operator
  - iv. no induction possible if repressor mutated so that inducer (allolactose) cannot bind
- Trp operon attenuation: deletions in leader sequence lead to increased production of mRNA for the five functional genes in the trp operon

RNA synthesis (transcription)

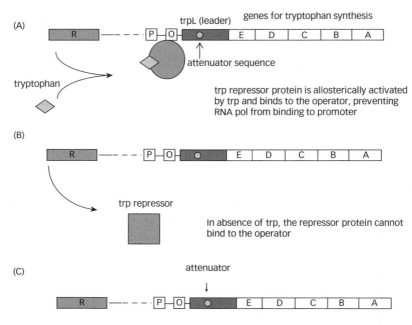

(A)

trpL (leader)   genes for tryptophan synthesis

attenuator sequence

tryptophan

trp repressor protein is allosterically activated
by trp and binds to the operator, preventing
RNA pol from binding to promoter

(B)

trp repressor

In absence of trp, the repressor protein cannot
bind to the operator

(C)

attenuator

RNA pol transcribes attenuator sequence which contains 2 adjacent codons for trp.
Ribosome follows behind RNA pol, translating the nascent mRNA (possible as
transcription and translation both occur in cytoplasm). Two possible outcomes:

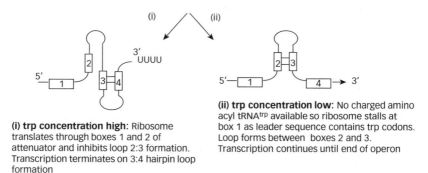

(i)

(ii)

(i) trp concentration high: Ribosome
translates through boxes 1 and 2 of
attenuator and inhibits loop 2:3 formation.
Transcription terminates on 3:4 hairpin loop
formation

(ii) trp concentration low: No charged amino
acyl tRNA^{trp} available so ribosome stalls at
box 1 as leader sequence contains trp codons.
Loop forms between boxes 2 and 3.
Transcription continues until end of operon

Figure 4.12 The trp operon. (A) The trp operon is negatively regulated by the trp
repressor which binds to the operator sequence in the presence of the co-repressor,
tryptophan, and prevents *RNA pol* from initiating transcription. (B) Relief of repression
in absence of tryptophan. When cells lack tryptophan, the trp repressor protein cannot
bind to the operator and the genes of the operon encoding enzymes required for
biosynthesis of tryptophan can be transcribed by *RNA pol* which now binds to the
promoter. (Note that trp allosterically activates the trp repressor protein unlike allolactose which allosterically inhibits the lac repressor.) (C) Control of transcription by
attenuation of the trp operon. Transcription is initiated and a ribosome starts translating
the attenuator peptide on the nascent RNA, following the *RNA pol*. The attenuator
peptide encodes a number of trp residues. If trp is abundant, the ribosome closely
follows RNA pol and sits over box 2, preventing 2 and 3 forming a hairpin. Instead, 3 and
4 form a hairpin and create a transcription terminator. If trp levels are low the ribosome
stalls on box 1. A hairpin between boxes 2 and 3 forms and transcription continues.

- ChIP on chip for *RNAP* shows which genes are actively transcribed
  - ➜ *see 9.11 Chromatin immunoprecipitation (ChIP) (p. 257) for details*
- RNA levels for specific genes can be measured by northern blotting or RT-PCR
  - ➜ *see 9.8 RNA manipulation and analysis (pp. 244–245) for details*

Typical eukaryotic gene structure

- Genes encoding proteins (i.e. transcribed into mRNA) are usually monocistronic, i.e. the transcript produced from that gene (mRNA) is translated into a single polypeptide (see Figure 4.13)
- Genes encoding rRNAs are polycistronic (18S, 5.8S, and 28S rRNA transcribed as single transcript which is then cleaved)
- Basal promoter elements of *RNA pol II*-transcribed genes are similar to prokaryotes:
  - i. TATA box (−25, in ~25% genes; important for locating *RNAP II* to transcription start site at +1)
  - ii. CAAT box (−70 to −80, enhances efficiency of transcription)
- Eukaryotic promoters are **modular**, i.e. comprise several/many different regulatory elements (also called 'boxes') allowing both positive and negative control of transcription, e.g. E-box (CACGTG) binds basic helix–loop–helix (bHLH) positive regulatory transcription factors such as c-myc
- Main regulatory boxes are located at specific distances from transcription start site and must be in correct orientation with respect to gene sequence
- **Enhancers** also exist—can be kb away from gene on either strand and in either orientation
- Three major classes of genes, each with different control elements and transcribed by different *RNA polymerases*:
  - i. *RNA pol* I genes: rRNA
  - ii. *RNA pol* II genes: mRNA (for protein), miRNAs (regulatory)
  - iii. *RNA pol* III genes: tRNA
- Many genes encoding proteins have <u>exons</u> (<u>ex</u>pressed regions) separated by <u>introns</u> (<u>in</u>tervening sequences): both are transcribed into pre-mRNA, then introns are removed by **splicing** to form mature mRNA

Regulation of initiation of mRNA transcription in eukaryotes

- Major regulatory step at the initiation of transcription through proteins binding to the modular promoter
- Core promoter (TATA and CAAT boxes) binds *RNA pol II* initiation complex and confers basal transcription levels
- Multiple general transcription factors (GTFs, especially TFIIA, B, D, E, F, and H) bind either to DNA at promoter or directly to *RNA pol II* in a large pre-initiation complex

RNA synthesis (transcription)

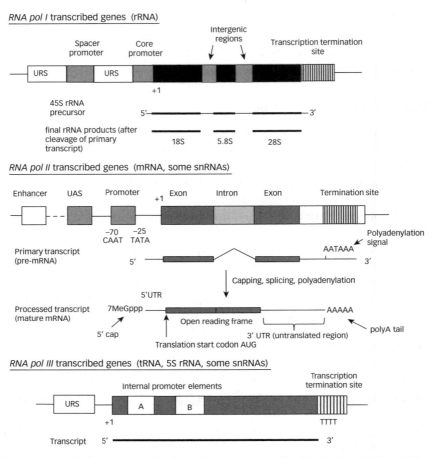

RNA pol I transcribed genes (rRNA)

RNA pol II transcribed genes (mRNA, some snRNAs)

RNA pol III transcribed genes (tRNA, 5S rRNA, some snRNAs)

Figure 4.13 Typical structures of eukaryotic genes transcribed by *RNA pols I, II* and *III*. The promoters and gene structures of eukaryotic genes differ depending on the *RNA pol* that transcribes them. URS = upstream regulatory sequence, UAS = upstream activating sequence. Some *RNA pol II* transcribed genes have a TATA box at –25 to which *RNA pol II* binds and CAAT box at –70 important for transcription factor binding.

- Gene-specific transcription factors that bind to upstream activating sequences (UAS) or enhancer elements positively regulate binding of *RNA pol II* initiation complex to promoter:
    - i. UAS are short DNA sequence elements found close to the transcription initiation site
    - ii. enhancer elements are DNA sequences that act at a distance from the initiation site and act independently of their orientation and position
- Transcription factors can be:
    - i. constitutively active, e.g. TATA binding protein (TBP)
    - ii. regulated by binding an allosteric activator, e.g. steroid hormone receptors (oestrogen receptor)

iii. regulated by phosphorylation, e.g. cAMP response element binding protein (CREB)

- Positive transcription factors recruit histone modifying enzymes, e.g. *histone acetylases* and chromatin remodelling complexes
- The '**histone code**' is important in regulating access of *RNAP* and transcription factor (TFs) to the promoter and so regulates transcription:
    i. histone acetylation increases transcription by 'loosening' chromatin
    ii. histone modifications associated with heterochromatin (e.g. methylation of lysine-9 on histone H3) prevent *RNA polymerase* access and hence block transcription
    iii. NB not all histone methylation negatively regulates transcription
- Nucleosome positioning is also important
- Silencer elements in DNA bind negative regulators of transcriptional initiation, e.g. *histone deacetylases*
- *RNAP II* is subject to phosphorylation on its C-terminal domain (CTD) in a series of heptad repeats: S5 phosphorylation by transcription factor TFIIH marks the switch from initiation to elongation; S2 is then phosphorylated by *Cdk9*

Eukaryotic mRNA processing

- Primary transcript (hnRNA or pre-mRNA) is often larger than final mRNA product due to presence of introns, e.g. dystrophin gene primary transcript 2,400 kb (2.4 Mb), mature mRNA 14 kb
- mRNA is capped with 7-methylguanine (7-MeG cap) at 5′ end linked by 5′–5′ phosphodiester bond (cap is important in translation of the mRNA)
- Introns removed by splicing in **spliceosome** in nucleus (see Figure 4.14)
- Transcript is cleaved and polyadenylated at 3′ end (poly(A) tail)—helps to stabilize the transcript
- mRNA consists of 5′ untranslated region (5′ UTR), protein coding region (contiguous exons) and 3′ untranslated region (3′ UTR)
- Capping, splicing and cleavage/polyadenylation happen during transcription and require *RNA pol II*
- Export of mRNA out of the nucleus is mediated by cap-binding proteins, i.e. is dependent on correct processing of primary transcript. Note only about 10% of primary transcripts are exported from the nucleus in humans

Revision tip

Don't get mixed up between intergenic regions, which are regions of DNA *between* genes, and introns (intervening sequences) that are found between exons *within* a gene.

RNA synthesis (transcription)

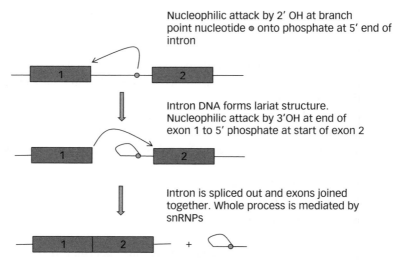

Nucleophilic attack by 2' OH at branch point nucleotide ● onto phosphate at 5' end of intron

Intron DNA forms lariat structure. Nucleophilic attack by 3'OH at end of exon 1 to 5' phosphate at start of exon 2

Intron is spliced out and exons joined together. Whole process is mediated by snRNPs

Figure 4.14 Intron splicing on mRNA processing in eukaryotes.

Evidence for regulation of mRNA transcription in eukaryotes

- Promoter insertion/deletion studies (usually coupled to 'reporter gene' with easily assayed protein product, e.g. *firefly luciferase* or *β-gal*) reveal requirement for positive DNA regulatory elements and the importance of their spacing relative to +1 transcription start site
- X-ray crystallography shows how TFs recruit *RNAP II* so that active site lies over +1 on template DNA
- Protection from *DNase I* cleavage reveals presence of proteins bound to promoters, enhancers, etc.
- Loss of gene expression is associated with UAS or transcription factor mutations in yeast
- Many purified transcription factor complexes possess histone acetylation activity, consistent with a role in activation of transcription
- *RNAP II* CTD mutations differentially affect initiation, elongation, or mRNA processing
- Table 4.3 compares transcription in prokaryotes and eukaryotes

Looking for extra marks?

MicroRNAs (miRNAs) are important in regulating eukaryotic gene expression by targeting cognate mRNAs for degradation through the RNA silencing pathway. They are transcribed by *RNA pol II* as longer precursors which form hairpin loops and are cleaved by a nuclease (*dicer*) to generate short (~22 nt) double-stranded RNA. One strand is incorporated into RISC (RNA-induced silencing complex); this recruits mRNAs with complementary sequence into RISC, where they are then cleaved and so 'silenced'.

Check your understanding

Describe the role of transcriptional repressor proteins in regulating prokaryotic gene expression. (*Hint: remember repressor proteins can be involved in regulating inducible (lac) as well as repressible (trp) operons.*)

What are the key similarities and differences between regulation of transcription in prokaryotes and eukaryotes? (*Hint: consider promoter structure, promoter recognition, key examples of regulation, attenuation in prokaryotes.*)

Feature	Prokaryotes	Eukaryotes
RNA polymerase	RNA pol ($\alpha_2\beta\beta'\sigma\omega$)	RNA pol I (28, 5.8, and 18 srRNA) RNA pol II (mRNA and miRNA) RNA pol III (tRNA and 5S rRNA)
Transcription start	+1	+1
Basal promoters (protein encoding genes)	−10 TATA (TAATAT) −35	−20 TATA −70 CAAT
Negative regulatory sequences	operators attenuators	URS (negative) silencers
Positive regulatory sequences	e.g. Crp binding site	UAS enhancers
Recognition of promoter elements	σ subunit	general transcription factors (GTFs) (three classes) TF$_I$ for RNA pol I TF$_{II}$ for RNA pol II TF$_{III}$ for RNA pol III plus specific transcription factors, e.g. c-Myc, HSTF, etc.
Initiation→elongation switch	σ dissociates	C-terminal domain (CTD) of RNA pol II is phosphorylated
Elongation	phosphodiester bond formation	phosphodiester bond formation
Termination	ρ-dependent ρ-independent	hairpin loop formation transcription barriers
Regulation	genes usually 'on' positive control, e.g. Crp protein negative control i. induction ii. repression iii. attenuation	genes usually 'off' positive control by transcription factors negative regulation: i. repression/heterochromatin silencing ii. squelching iii. RNA$_i$ silencing
mRNA processing	−	5'–5' 7Me-G cap at 5' end polyA tail at 3' end splicing to remove introns
rRNA processing	cleavage of large transcript	cleavage of large transcript
tRNA processing	large number of modified bases	large number of modified bases

Table 4.3 Comparison of transcription in prokaryotes and eukaryotes

4.3 PROTEIN SYNTHESIS (TRANSLATION)

The synthesis of proteins by polymerization of amino acids using information encoded in mRNA template using tRNA adaptors, taking place within the ribosome.

Protein synthesis (translation)

- Triplet genetic code—three bases in mRNA code for one amino acid, but redundant since 64 (4^3) possible codons and only 20 commonly occurring amino acids
- Three major steps: initiation, elongation, and termination
- Sequence in 5′ region of mRNA <u>upstream</u> of initiation codon is initially bound by **ribosome**
- Translation starts at initiation codon—usually AUG, sometimes GUG in prokaryotes
- Polypeptide synthesized from amino (N) to carboxyl (C) terminus (5′→3′ of mRNA)
- Amino acids are covalently bonded to specific tRNA molecules which act as adaptors
- Charged tRNA recognizes a triplet in mRNA by complementary base pairing between anticodon on tRNA and codon on mRNA
- Peptide bonds formed between amino acids carried by the tRNAs, catalysed by an RNA enzyme (*ribozyme*) in the ribosome
- Translation stops at one of three specific stop codons
- 3′ untranslated regions (UTRs) exist downstream from stop codons
- Translation uses energy to:
    i. charge tRNA with amino acid (ATP)
    ii. remove initiation factors to allow ribosome assembly (GTP)
    iii. translocate ribosome three bases along mRNA (GTP)
    iv. NB no nucleotide hydrolysis directly involved in peptide bond formation
- Essentially similar in prokaryotes and eukaryotes though details differ
- Eukaryotic translation takes place in cytosol or on rough endoplasmic reticulum; in prokaryotes, translation occurs in cytoplasm on nascent mRNA while it is still being transcribed

Genetic code

- Triplet—three bases in mRNA (**codon**) code for one amino acid
- Redundant:
    i. four possible bases in three positions, i.e. $4^3 = 64$ possible codons
    ii. three stop codons (UGA, UAG, UAA)
    iii. therefore 61 codons for 20 amino acids and most amino acids are encoded by more than one codon
- Redundancy is mainly in third base of codon = wobble base (e.g. glycine GGX where X can be G, A, U, or C)
- Start codon = AUG (methionine), sometimes GUG in prokaryotes
- Genetic code is almost universal (very rare exceptions, e.g. some organelle genomes)

Key evidence for genetic code

- Comparison of sequences from a variety of organisms shows universality
- Code deduced from *in vitro* translation of synthetic RNAs with repeating trinucleotides and a single radiolabelled **aminoacyl-tRNA**, e.g. polyU, directs the polymerization of phenylalanine (codon = UUU)
- Altering a single base changes only one amino acid, i.e. the code is non-overlapping
- A trinucleotide will bind to a ribosome and promote the binding of a single tRNA species—all 64 trinucleotides were synthesized and each one tested to see if it would promote binding of a tRNA charged with a ^{14}C-labelled amino acid
- Nucleotide polymers with defined sequence can be translated *in vitro* to generate defined proteins

tRNAs

- Adaptor molecules:
  - i. anti-codon (three bases within anticodon loop) is complementary to codon on mRNA
  - ii. 3′ terminus is covalently bound to specific amino acid
- Bring amino acid to correct codon on mRNA via base pairing between the tRNA anticodon and mRNA codon; ensure that the correct amino acid is added to growing polypeptide chain
- Initiator tRNA binds into partial P site of small subunit of ribosome (prior to assembly of whole ribosome):
  - i. fmet-tRNA$_i$ in prokaryotes (carries *N*-formyl methionine)
  - ii. met-tRNA$_i$ in eukaryotes
- All other charged aa-tRNA bind into A site of complete ribosome
- Contain many unusual nucleotides (modified from standard nucleotides after initial transcription of tRNA)
- Specificity of tRNA for codon lies mainly in pairing to first two nucleotides of codon so one tRNA can recognize more than one codon, varying in the wobble position
- Many tRNAs contain a modified base (e.g. inosine) in first base of anticodon, corresponding to third, or 'wobble' base in codon:
  - i. can base pair with more than one nucleotide, i.e. the wobble hypothesis
  - ii. reduces number of tRNAs required (20 < tRNAs < 64, i.e. sufficient to carry all naturally occurring amino acids, but fewer than total possible codons)
- 2D structure = cloverleaf; 3D structure = banana shape (Figure 4.15)
- Sequence of many tRNAs is known and conserved base-pairing gives a conserved structure
- 3D crystal structure of some tRNAs known and can be used to model structure of other tRNAs

Protein synthesis (translation)

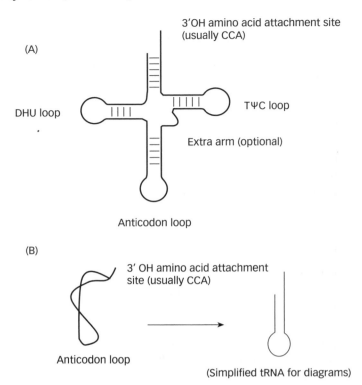

Figure 4.15 Structural features of tRNAs. tRNAs are 73–93 nucleotides long and contain up to 20% unusual bases (e.g. inosine), modified post-transcriptionally. The amino acid is attached to the 3' end, usually CCA. The 5' end is phosphorylated, usually pG. (A) Note the significant regions of complementarity resulting in hairpin loop formation. The classical cloverleaf is actually the 2D representation of tRNA. (B) The 3D structure is more like an upside down 'L' shape or banana and can be drawn very simply as shown.

Aminoacyl tRNA synthetases

- Enzymes responsible for attaching amino acid to correct tRNA
- ≥20 different enzymes (i.e. one per tRNA, ≥1 per amino acid)
- Main source of specificity in protein synthesis:
    i. essential to add correct amino acid onto appropriate tRNA as ribosome does not check whether amino acid and tRNA anticodon match
    ii. experimental modification of amino acid R group (e.g. cys → ala) after joining to tRNA leads to incorporation of modified incorrect amino acid into polypeptide chain by ribosome
- Recognize:
    i. amino acid via properties of R chain (e.g. size, hydrophobicity)
    ii. tRNA via anticodon loop and/or 3' acceptor stem

- Charge tRNAs with amino acid (aa) in two step reaction requiring ATP hydrolysis:

$$aa + ATP \longrightarrow aa - AMP + PP_i \xrightarrow{\text{tRNA}} aa - tRNA$$
$$\downarrow$$
$$2P_i$$

- Proofreading activity removes incorrectly attached amino acid. Editing occurs at the level of either:
    i. aminoacyl-AMP, e.g. isoleucine
    ii. aminoacyl-tRNA, e.g. valine
- Proof-reading uses energy from ATP hydrolysis, i.e. costly
- Crystal structure of *aa-tRNA synthetase* confirms mechanism–activation step and transfer to tRNA on different domains
- Altering the sequence of the anticodon loop allows determination of the residues required for specificity of interaction between tRNA and *synthetase*

Ribosomes

- Large protein–rRNA complexes (see Table 4.4 for ribosome subunits and sizes)
- Catalyse protein synthesis (see 1.2 proteins (p. 4) for protein structure)
- Interact with mRNA and charged tRNA^{aa}s, aligning adjacent amino acids for peptide bond formation according to base sequence on mRNA
- Sites of tRNA^{aa} binding:
    i. P (peptidyl) site: binds tRNA coupled to nascent polypeptide, or initiator tRNA^{met} in partial P site
    ii. A (amino acyl) site: binds all other incoming tRNA^{aa}
    iii. E (exit) site—prokaryotes only: uncharged tRNAs exit from ribosome via this site
- Two major enzymatic activities:
    i. *peptidyl transferase*—an RNA enzyme (ribozyme) which catalyses peptide bond formation
    ii. *translocase*—moves ribosome three bases along the mRNA and requires energy from GTP hydrolysis
- rRNA in prokaryotes has various roles (possibly similar roles in eukaryotes):
    i. scaffold for ribosome structure
    ii. catalytic activity (*peptidyl transferase*)
    iii. 16S rRNA aligns with **Shine–Dalgarno** region upstream of initiation codon of mRNA to place start codon in partial P site
    iv. 23S rRNA interacts with tRNA
- Several ribosomes can translate a single mRNA at once = polyribosome

Protein synthesis (translation)

Subunit/ component	Prokaryotes	Eukaryotes	Mitochondria	Chloroplasts
Overall sedimentation coefficient	70S	80S	55–77S	70S
Large	50S	60S	39–60S	46–54S
rRNAs	23S, 5S	28S, 7S, 5.8S	16–25S, 5S	23S, 5S, 4.5S
proteins	~34	~40		
Small	30S	40S	28–40S	28–35S
rRNAs	16S	18S	12–14S	16S
proteins	~21	~30		
Total mass	2,500 kDa	4,500 kDa	3,200–4,500 kDa	2,500–3,300 kDa

Table 4.4 Comparison of ribosomes

- Ribosomal proteins are not essential for ribosome function but increase efficiency of translation
- Different sizes in prokaryotes, eukaryotes, and organelles (Table 4.4)
- Organelle ribosomes more closely resemble prokaryotic than eukaryotic—support for **endosymbiotic theory**

Key evidence for ribosomes

- Size of subunits and intact ribosome determined by sedimentation, which depends on both mass and shape (hence 30S + 50S = 70S for prokaryotes)
- Detectable as single subunits and dimeric complexes as well as polyribosomes by electron microscopy (EM) and analytical gradient centrifugation
- Location of proteins in 30S subunit determined by neutron diffraction analysis
- Can assemble *in vitro* from component parts; omitting individual components allows their function to be determined
- Ribosomal proteins alone will not form complexes, i.e. require rRNA scaffold
- Cleavage of 16S rRNA blocks protein synthesis in prokaryotes

Steps in translation in prokaryotes

1. Initiation
- 30S subunit associates with initiation factors IF1, 2, and 3 (Figure 4.16)
- Shine–Dalgarno sequence of mRNA (purine-rich sequence of 3–9 bases, approximately ten bases upstream of start codon AUG) associates with 16S rRNA of 30S ribosome subunit by base pairing
- **Reading frame** of protein synthesis defined by this first interaction of rRNA with mRNA
- Initiator tRNA is different from tRNAs that carry methionine encoded by internal AUG codons: fmet-tRNA$_i^{fmet}$ (initiator tRNA) is the only charged tRNA that can enter partial P site of 30S subunit

Protein synthesis (translation)

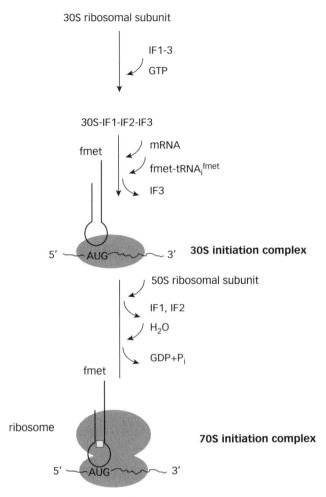

Figure 4.16 Initiation of translation in prokaryotes. Aided by initiation factors (IFs), the small ribosomal subunit recognizes the Shine–Dalgarno sequence on the mRNA and recruits the initiating aa-tRNA (fmet-tRNA$_i^{fmet}$) to the partial P site. The large ribosomal subunit joins after dissociation of the IFs.

- Anticodon CAU ($5' \rightarrow 3'$) of initiator tRNA base pairs with start codon AUG ($5' \rightarrow 3'$) on mRNA
- GTP is hydrolysed (catalysed by ribosomal proteins L7 and L12) to release **initiation factors** IF1 and IF2
- 50S subunit associates to form 70S complex; fmet-tRNA$_i^{fmet}$ now in complete P site

2. Elongation

- tRNA carrying second amino acid (aa2-tRNA) enters A site (Figure 4.17)
- Anticodon of tRNA base pairs with codon on mRNA

Protein synthesis (translation)

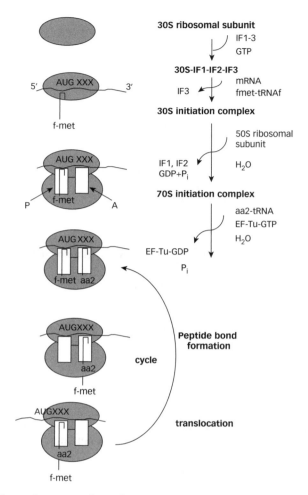

Figure 4.17 Elongation stage of translation in prokaryotes (E site not shown).

- Aided by **elongation factor** EF-Tu in GTP-bound form. EF-Tu GTPase activity hydrolyses bound (GTP → GDP + P$_i$):
  i. conformation of EF-Tu alters—only correctly base-paired aa-tRNA remains associated with mRNA and ribosome (specificity step)
  ii. EF-Tu:GDP released from ribosome
  iii. EF-Tu:GTP regenerated by action of EF-Ts (Figure 4.18)
- Peptide bond formed by *peptidyl transferase* (an RNA enzyme)
- mRNA moves three bases with respect to ribosome so that next codon is now positioned in A site—mediated by *translocase*
- Uncharged tRNA leaves P site and moves to E site (not shown in Figure 4.17)
- Peptidyl-tRNA moves from A to P site
- Cycle of elongation is repeated

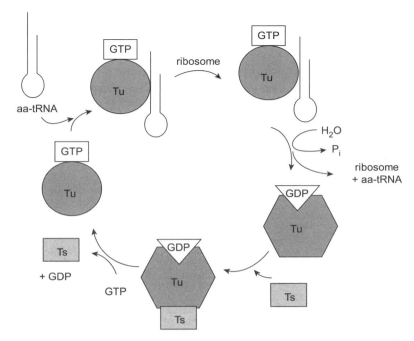

Figure 4.18 Recycling of elongation factor EF-Tu in translation. EF-Tu is required for association of the correct aa-tRNA in the A site of the ribosome. When the correct aa-tRNA is bound, the GTP bound to EF-Tu is hydrolysed to GDP to promote EF-Tu leaving the complex. EF-Ts is required to catalyse the exchange of the GDP for GTP to allow the EF-Tu to function again. Without this, the cell would rapidly run out of EF-Tu-GTP and protein synthesis would halt.

3. Termination

- Three stop codons: UAA, UAG, UGA
- No tRNA normally exists with anti-codon complementary to stop codons (except mutant 'suppressor' tRNAs)
- Stop codons in A site of ribosome recognized by **release factors**:
    - i.   RF1 recognizes UAA and UAG
    - ii. RF2 recognizes UAA and UGA
- Specificity of *peptidyl transferase* is altered so that it hydrolyses the bond between the amino acid and the tRNA, i.e. water is final acceptor
- Polypeptide and mRNA are released from ribosome
- 30S and 50S ribosomal subunits dissociate and are kept apart in cytoplasm by IFs

Evidence for steps in protein synthesis

- Shine–Dalgarno sequence identified by comparison of many mRNA sequences
- Mutations in Shine–Dalgarno sequence block initiation—compensating mutations in 16S rRNA, which restore base pairing, allow initiation

Protein synthesis (translation)

- Elongation requires the presence of EF-Ts to recycle the EF-Tu
- Elongation is blocked in the presence of non-hydrolysable analogues of GTP, e.g. GTPγS
- All open reading frames end with one or more stop codon
- Structure of ribosome during various steps in translation determined by X-ray crystallography
- Antibiotics can be used to probe protein synthesis at various stages (see Table 4.5)

Antibiotic	Target	Effect
Chloramphenicol	prokaryotic, mitochondrial, and chloroplast *peptidyl transferase*	blocks peptide bond formation
Cycloheximide	eukaryotic 60S subunit	blocks EF-2 dependent translocation
Erythromycin	prokaryotic 50S subunit (and mitochondrial in, for example, yeast).	blocks translocation
Puromycin	structurally resembles 3′ end of aminoacyl-tRNA so competes with aa-tRNAs for A site on 50S subunit of prokaryotic ribosome.	*peptidyl transferase* transfers nascent polypeptide to puromycin via non-cleavable bond, resulting in termination of translation
Streptomycin	S12 protein on 30S subunit of prokaryotic ribosome.	inhibits initiation by blocking fmet-tRNA interaction with ribosome
Tetracycline	*peptidyl transferase* of 50S subunit of prokaryotic ribosome	blocks peptide bond formation

Table 4.5 Targets of antibiotics that act on protein synthesis

Revision tip

The stages in translation are complex and hard to visualize. Look at animations produced by the Nobel prize winning Yonath group (Israel) to get a 'feel' of the process (e.g. http://www.youtube.com/watch?v=Jml8CFBWcDs).

Examples of translational regulation

Prokaryotes: presence or absence of Shine–Dalgarno sequence
- Phage λ repressor:
    i. mRNA transcribed from P_R has Shine–Dalgarno sequence and is efficiently translated (leads to high levels λ repressor protein to establish state of lysogeny)
    ii. mRNA transcribed from P_{RM} lacks Shine–Dalgarno sequence and is inefficiently translated (low levels of λ repressor protein are required to maintain the lysogenic state)

Eukaryotes can regulate translation by phosphorylation of initiation or elongation factors:
- eIF-2 phosphorylation to eIF-2P in response to interferon upon viral infection:
    i. eIF-2P forms a stable complex with guanine nucleotide exchange factor
    ii. prevents regeneration of eIF-2:GTP
    iii. inhibits formation of initiation complex

iv. important pathway in blocking viral proliferation by shutting down protein synthesis in infected cells

- Phosphorylation of eIF4B (by *S6 kinase*) promotes formation of the translation pre-initiation complex
- eIF4E is sequestered by 4E-BP1; phosphorylation of 4E-BP1 (by *mTOR*) releases eIF4E which can then recruit capped mRNAs to the ribosome
- Phosphorylation of eIF4A alters in response to starvation in eukaryotic cells so starving cells downregulate protein synthesis
- Table 4.6 compares translation in prokaryotes and eukaryotes

Component	Prokaryotes	Eukaryotes
Ribosome	30S + 50S = 70S	40S + 60S = 80S
Start codon	AUG or GUG	AUG
Initiator amino acid	*N*-formyl methionine usually cleaved from mature protein	methionine usually retained in mature protein
mRNAs	often polycistronic	monocistronic
Ribosome binding site (rbs) on mRNA	Shine–Dalgarno sequence (S-D) on mRNA	5′ cap and cap-binding protein on mRNA Kozak consensus sequence for preferred nucleotides surrounding start codon
rbs recognized by	16S rRNA	40S ribosome subunit
Determination of start codon	placed in partial P site by Shine–Dalgarno:16S rRNA interaction	scan from 5′ cap
Initiation factors	three IFs	≥9 eIFs
Elongation factors	two (EF-Tu and EF-TS)	many eEFs
Termination factors	RF1, 2, and 3	only one
Regulation of initiation	GTP hydrolysis → loss of IF2	phosphorylation → activation or inactivation of IFs

Table 4.6 Comparison of prokaryotic and eukaryotic translation

Looking for extra marks?

Note major differences between prokaryotic and eukaryotic translation:
- No sequence equivalent to the prokaryotic Shine–Dalgarno sequence is recognizable in eukaryotes
- Altering the 5′ untranslated sequence to contain an AUG shifts the point of translational initiation in eukaryotes.
- Many antibiotics specifically inhibit prokaryotic but not eukaryotic translation (Table 4.5)

 Check your understanding

What factors ensure that the correct amino acid is added to the elongating peptide chain? (*Hint: consider amino acyl tRNA synthetases as well as codon:anticodon base pairing, and GTP hydrolysis on EF-Tu.*)

continued

> Using labelled diagrams, trace the key steps in prokaryotic protein synthesis. (*Hint: make sure you draw very simple diagrams with all relevant parts labelled. You can add lots of information by annotating the diagrams, e.g. with boxes of text pointing to the relevant part.*)

Comparison of DNA replication, transcription and translation

- DNA replication, RNA transcription, and protein synthesis (translation) all occur via three distinct steps of initiation, elongation, and termination (Table 4.7)
- Specific start and stop sites are required on each template
- Multi-subunit enzyme complexes are required for the synthesis of DNA, RNA, and protein
- All processes are regulated in order to control genome stability and gene expression (see Figure 4.1)

Characteristic	Replication	Transcription	Translation
Template	DNA (both strands)	DNA (one strand)	mRNA
Product	DNA (double-stranded) one copy per cell—very stable	RNA multiple copies per gene, transient with relatively short half-life	protein multiple copies per mRNA, transient though half-lives vary considerably
Building blocks	dNTPS	rNTPS	amino acids
Initiation site defined by	origin	promoter	start codon plus ribosome binding site
Elongation	phosphodiester bond formation	phosphodiester bond formation	peptide bond formation
Mediated by	*DNA polymerases*	*RNA polymerases*	ribosome + tRNAs
Termination	TUS/Ter in *E. coli*	ρ-dependent and independent sites in *E. coli* hairpin loops in eukaryotes	stop codons
Chemical inhibitors	aphidicolin (eukaryotes) nalidixic acid (prokaryotes)	α-amanitin (eukaryotes) rifampicin (prokaryotes)	cycloheximide (eukaryotes) tetracycline (prokaryotes)

Table 4.7 Comparison of replication, transcription, and translation

4.4 DNA REPAIR

The process by which damaged DNA is corrected to maintain genomic integrity.

Key concepts in DNA repair

- DNA damage causes **mutations** (permanent changes to DNA) if not repaired (Figure 4.19)
- Some mutations are deleterious so stability of genes and genome depends critically on DNA repair

continued

- Failure of DNA repair can lead to cancer and other genome instability syndromes (vertebrates)
- Spontaneous damage includes deamination (e.g. cytosine to uracil), hydrolysis of bases or sugar-phosphate backbone, and oxidation of bases, e.g. 8-oxo-G
- Environmental damage includes ultraviolet (UV) light, ionizing radiation, and chemicals that bind to or modify DNA, e.g. methylmethonyl sulphate (MMS), cisplatin
- Damage is sensed and signalled through a DNA damage response signalling pathway (DDR in eukaryotes, SOS in prokaryotes), so cells synthesize and/or activate DNA repair proteins
- Three ways of dealing with damage:
    - i. reverse directly
    - ii. repair
    - iii. tolerate (if damage is too great, cells die by programmed cell death— **apoptosis**)
- Type of repair depends on nature of damage:
    - i. DNA damaged on one strand only is repaired according to sequence in undamaged complementary strand
    - ii. DNA with double strand damage can be repaired using information on undamaged DNA duplex, e.g. **sister chromatid** or **homologous chromosome** (in diploids)
- Major classes of DNA damage reversal/repair:
    - i. direct reversal (single damaged base)
    - ii. base excision repair (BER) (single damaged base)
    - iii. nucleotide excision repair (NER) (several damaged bases on same strand)
    - iv. mismatch repair (MMR) (replication errors)
    - v. non-homologous end joining (NHEJ) (double strand breaks)
    - vi. homologous recombination (HR)) (double strand breaks)
    - vii. interstrand crosslink repair (ICL repair) (cross-linked DNA strands)
- Tolerance mechanisms involve translesion *DNA polymerases* that that can use the damaged template for DNA replication—error-prone pathway
- Repair is very efficient, between 10^4 and 10^6 spontaneous lesions repaired per cell per generation

Direct reversal of damage

- Photoreactivation:
    - i. repairs pyrimidine dimers caused by exposure of DNA to UV light
    - ii. uses energy from visible light (300–500 nm)
    - iii. enzyme *photolyase* reverses formation of cyclobutane ring between adjacent pyrimidines on same DNA strand

DNA repair

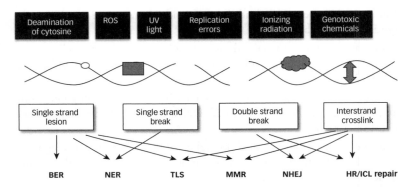

Figure 4.19 DNA damaging agents, types of DNA damage, and the processes used to repair them. ROS = reactive oxygen species.

- Reversal of alkylation of bases:
    - i. uses sacrificial enzyme *methyl guanine methyl transferase* (*MGMT*) in eukaryotes, equivalent to *ogt* for alkylated G and *ada* for alkylated C (and T) in bacteria
    - ii. enzymes become methylated stoichiometrically so can only function once
    - iii. methylated *ada* acts as a transcription factor to induce production of more *ada*

Base excision repair (BER)

- Repairs single damaged base
- Very rapid
- Steps in BER (Figure 4.20):
    - i. recognition of damaged base, e.g. presence of uracil in DNA from spontaneous deamination of cytosine
    - ii. specific nuclease removes damaged base from DNA by breaking glycosidic bond releasing base and leaving apyrimidinic (AP) site but intact sugar–phosphate backbone (e.g. *uracil DNA glycosylase* removes uracil)
    - iii. *AP endonuclease* cleaves backbone at AP site
    - iv. gap filled by *DNA polymerase*
    - v. nick in backbone sealed by *DNA ligase*
- Two pathways in eukaryotes:
    - i. major pathway (>70% BER): single damaged base replaced by *DNA pol β*
    - ii. minor pathway (30% BER): synthesis carried out by *DNA pol β* or *pol δ/ε* with additional nuclease *FEN1* which cleaves flap 5′ to AP site and 6–14 nucleotides removed; requires also PCNA

Nucleotide excision repair (NER)

- Repairs damage on single strand of DNA, e.g. pyrimidine dimers and 6–4 photoproducts caused by UV light

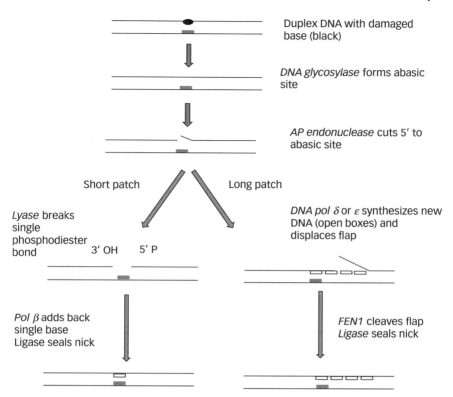

Duplex DNA with damaged base (black)

DNA glycosylase forms abasic site

AP endonuclease cuts 5′ to abasic site

Short patch

Long patch

Lyase breaks single phosphodiester bond

3′ OH   5′ P

DNA pol δ or *ε* synthesizes new DNA (open boxes) and displaces flap

Pol β adds back single base
Ligase seals nick

FEN1 cleaves flap
Ligase seals nick

Figure 4.20 Base excision repair. A single damaged base on one strand of the DNA is removed by cleaving the glycosidic bond, leaving an intact backbone but an abasic site. The backbone is nicked 5′ to this site, and either short patch or long patch BER takes place. (Note it is easier to depict BER if you do not show H bonds between the base pairs, but it is important to remember that they are present and hold the DNA together).

- May be slower than BER—up to 24 hours for complete repair after exposure of cell to UV light
- Inability to repair UV damage results in heritable skin cancer syndromes in man, e.g. xeroderma pigmentosum (XP) and Cockayne syndrome (CS)
- Human XP proteins act in similar way to bacterial *uvrABC exinuclease*
- Two types in eukaryotes:
    - i. global genome repair (GGR) repairs all of the genome equally
    - ii. transcription-coupled repair (TCR) preferentially repairs actively transcribed genes
- Steps in NER in mammals (Figure 4.21):
    - i. recognition of damage (human XPC, E, A, and/or CS proteins)
    - ii. DNA backbone cleaved by nucleases 3′ (*XPG*) and 5′ (*XPF* with ERCC1) of damage

   iii. approximately 12 nucleotides in *E. coli* or 27 nucleotides in eukaryotes removed between cleavage sites (requires *helicase* activity of XPB and XPD)

   iv. DNA complementary to undamaged strand resynthesized by *DNA pol I* in *E. coli*, and *DNA polymerase δ or ε* in eukaryotes (with PCNA)

   v. nicks in backbone sealed by *DNA ligase III*

Mismatch repair (MMR)

- Repairs errors caused during DNA replication or **base tautomerization** which result in non-complementary (i.e. mismatched) bases paired in duplex
- Detects distortion of double helix due to mismatched bases
- Several nucleotides or longer regions up to 1 kb may be removed
- Steps in MMR in bacteria:
  - i. recognition of mismatch (helical distortion) by MutS dimer
  - ii. MutL dimer binds MutS-DNA complex and activates MutH which is bound at hemimethylated sites
  - iii. DNA is looped out from mismatch to nearest d(GATC)
  - iv. *MutH* nicks the daughter strand
  - v. *uvrD helicase* separates the strands
  - vi. *MutSHL* complex slides along the ss DNA
  - vii. exonuclease degrades the daughter ss DNA—if nick is 5′ of mismatch, *RecJ*, or *ExoVIII* are used; if nick 3′ end to the mismatch, *Exo1* is used

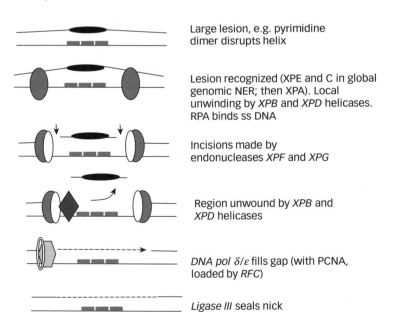

Large lesion, e.g. pyrimidine dimer disrupts helix

Lesion recognized (XPE and C in global genomic NER; then XPA). Local unwinding by *XPB* and *XPD* helicases. RPA binds ss DNA

Incisions made by endonucleases *XPF* and *XPG*

Region unwound by *XPB* and *XPD* helicases

DNA pol δ/ε fills gap (with PCNA, loaded by *RFC*)

Ligase III seals nick

Figure 4.21 Steps in nucleotide excision repair in eukaryotes.

viii. *DNA pol* III fills the gap using the parental strand as template

ix. *ligase* seals the nick in the backbone

x. *DNA methylase* methylates the newly synthesized daughter DNA

- Repair system must distinguish between parental strand and new strand of DNA to correct newly introduced error in new DNA strand rather than template base on parental strand:

  i. parental DNA strand in bacteria is identified by methylation of A residues in sequence GATC; *methylase* acts after replication so A residues in new DNA strand are not yet methylated

  ii. in eukaryotes, nascent DNA contains nicks (between adjacent Okazaki fragments prior to ligation), allowing repair machinery to distinguish intact parental from nascent lagging strand

  $\circledarrow$ *see Steps in DNA replication (pp. 81–86) for more information on Okazaki fragments*

- Mismatch repair proteins are also critical in minimizing illegitimate recombination between very similar DNA sequences throughout the genome

- Cells deficient in any of the Mut proteins have a greatly increased rate of spontaneous mutation—implicated in, for example, colon cancer in humans

Non-homologous end joining (NHEJ)

- Repairs double strand DNA breaks caused, for example, by ionizing radiation, when no intact duplex template is available, e.g. in G1 phase of cell cycle of diploid organisms

- Is also used in V(D)J recombination of antibody genes in B cells

- Steps in NHEJ and in eukaryotes:

  i. broken DNA ends at double strand break are recognized and bound by MRN complex, Ku heterodimer, and DNA-PK$_{CS}$

  ii. 'dirty' ends repaired, e.g. nuclease (possibly *Artemis*) trims ends

  iii. ends are joined by *DNA ligase IV*, XRCC4 and XLF

- Usually error-prone and mutagenic, i.e. results in loss of DNA sequence information

Homologous recombination (HR)

- Repairs double strand DNA breaks caused, for example, by ionizing radiation

- Requires intact homologous DNA region, e.g. on intact sister chromatid (so occurs in S phase or G2) or uses a homologous chromosome (meiosis)

- Steps in HR in eukaryotes (Figure 4.22):

- HR allows damaged/missing DNA sequence to be replaced in an error-free manner:

  i. strand resection at double strand break

  ii. invasion of 3′ overhang into homologous duplex (e.g. sister chromatid) to form a displacement or D loop

DNA repair

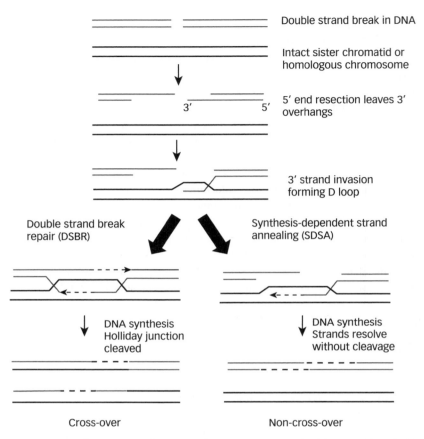

Figure 4.22 Homologous recombination. This process repairs double-strand breaks if an identical sister chromatid or homologous chromosome is available. SDSA is error-free and restores the genetic information. DSBR may result in cross-overs which can alter genetic information if between homologous (not identical) chromosomes. This is important for generating genetic variation during meiosis but can be deleterious if resulting from DNA damage. For clarity, the proteins involved are not shown. (Note: in exam diagrams, use different colours to differentiate between the various strands.)

    iii. synthesis of DNA from the 3′ end to replace missing sequence

    iv. then:

        a. formation of double **Holliday junction** (HJ), that can be resolved by cleavage at the cross-over points (cross-over can results in genetic alterations) or by *helicase*-mediated resolution ('flipping' back) of the strand (avoiding cross-over) or

        b. synthesis-dependent strand annealing (SDSA): the single HJ branch is migrated back (*helicase* activity) and the single nascent strand is joined back onto the break point

- Strand invasion is mediated by formation of a nucleoprotein filament using RecA (bacteria) or *Rad51* (eukaryotes). The protein:
  - i. coats ssDNA stoichiometrically
  - ii. scans intact duplex DNA (e.g. sister chromatid) for complementary regions, i.e. regions of homology
  - iii. catalyses strand assimilation reaction
  - iv. requires ATP hydrolysis to form D loop structure
- *recBCD* complex in *E. coli* has unwinding and nuclease activities that generate a ssDNA region to initiate strand invasion
- Mutants in recombination proteins are very sensitive to agents causing double strand DNA breaks such as ionizing radiation (e.g. recA⁻ *E. coli*, Bloom syndrome in humans, BRCA1 and 2 in human breast cancer)
- If homologous DNA region not available, error-prone repair may take place using any stretch of DNA as template

Looking for extra marks?

New 'synthetic lethality' cancer therapies use inhibitors of NHEJ in cells lacking HR pathway components (e.g. BRCA2-null breast cancer). Cells treated this way cannot repair double strand breaks and so die on treatment with agents that cause double strand breaks.

Interstrand crosslink repair (ICL repair)

- Repairs sites where both strands of DNA are joined by intercalating chemicals (e.g. mitomycin C, cisplatin, also ROS damage) which blocks DNA unwinding and hence inhibits transcription or replication
- Requires proteins of the Fanconi anaemia (FA) pathway—FA patients are hypersensitive to agents that cross-link DNA
- FANC protein core complex is recruited to damaged DNA, leading to modification of FANC D2 and FANC I proteins

SOS response (prokaryotes)

- Triggered by DNA damage in prokaryotes
- Results in increased synthesis of >15 repair proteins; transcription of these repair genes is usually blocked by lexA repressor in undamaged cells
- ssDNA resulting from damage is bound by recA protein, activating a protease activity of *recA*
- *recA* hydrolyses lexA, relieving transcriptional repression
- Repair genes, including *recA*, are now transcribed and translated
- Utilized by phage λ, as *recA* cleaves λ repressor protein and releases phage from lysogenic to lytic life cycle to 'escape' damaged bacterial cell

DNA repair

- Eukaryotic equivalent to bacterial SOS pathway
- Steps include:
    i. sense damage and activate signalling pathway (e.g. kinases *ATM/R*)
    ii. recruit mediators (53BP1, MDC1, BRCA2, etc.)
    iii. transduce signal (e.g. checkpoint kinases *Chk1* and *Chk2*)
    iv. effect response (e.g. tumour suppressor gene product p53)
- Phosphorylation of p53 stabilizes it (prevents its breakdown)
- Activated p53 acts as a transcription factor for proteins that halt the cell cycle (e.g. p21), promote DNA repair (e.g. Gadd45) or cause apoptosis (e.g. Bax, Puma)
- Table 4.8 compares types of DNA repair

 Check your understanding

Describe two pathways by which DNA damage can be repaired. (*Hint: make sure you can draw labelled diagrams for each of the pathways you have chosen and give named examples of what happens when they go wrong.*)

Compare and contrast DNA replication and transcription of RNA (*Hint: think about template, start sites, enzymes, bonds formed, and steps of initiation, elongation, and termination.*)

Compare the action of DNA polymerases in DNA replication and repair (*Hint: discuss pol α, δ, and ε in replication of iDNA, leading and lagging strands of processivity and importance of primer, vs. pol β, δ and ε in repair synthesis shorter product, use free 3' OH of DNA end as primer. Polarity, bonds formed, and mechanism of polymerization same in each.*)

Type of repair	Damage	Pathway steps	Diseases when deficient in repair
Reversal	alkylated bases or UV-damaged bases	• single step reversal of damage (alkylation reversed by transferring alkyl group to suicidal enzyme)	cancer if deficient in *MGMT*
BER	single base damage, e.g. 8-oxo-G, uracil in DNA	• removal of damaged base by cleaving glycosidic bond • cleavage of backbone 5' of AP site • replacement synthesis of the missing base • ligation of nick	*pol β* mutation found in 30% human cancers
NER	several bases damaged on same strand, e.g. UV products	• recognition of lesion • cut 5' of lesion • cut 3' of lesion • *helicase*-mediated removal of cut strand • DNA synthesis to fill gap • ligation	xeroderma pigmentosum (XP), Cockayne syndrome (CS), trichothiodystrophy (TTD)

Type of repair	Damage	Pathway steps	Diseases when deficient in repair
MMR	mismatched base(s)—replication error or base tautomerization	• mismatch recognized • DNA loops out to hemimethylated site • DNA nicked 3' or 5' of mismatch • *helicase* unwinds duplex • Mut complex slides on ss DNA • *nuclease* degrades ss daughter strand • DNA synthesis to fill gap • nick sealed by *ligase* • new DNA methylated	cancer, e.g. NPCC (colon cancer)
NHEJ	double strand break	• broken ends bound by protein • *nuclease*-mediated trimming of ends • join trimmed ends together	immune deficiency (SCID)
HR	double strand break	• recognize broken ends • cut away at break to leave 3' overhangs • *Rad51/RecA*-driven strand invasion • DNA synthesis using intact complementary strand as template • Holliday junction formed and HJ resolved by *helicase* or *nuclease* • or SDSA: non-cross-over pathway for error-free repair without changing genetic sequence	ataxia telangiectasia (AT), cancer (e.g. Bloom syndrome, breast cancer) premature ageing (Werner syndrome)
ICL repair	both strands of a duplex cross-linked (e.g. during cancer chemotherapy)	• recognize lesion • recruit FANC core complex • then can be repaired by HR, NER, translesion synthesis, etc.	Fanconi anaemia

Table 4.8 Comparison of types of DNA repair

4.5 DNA RECOMBINATION

Recombination is the formation of new gene arrangements by movement of sections of DNA. General recombination requires large regions of homology, while no homology is necessary for transposition. Recombination is important in generating genetic diversity, e.g. in eukaryotic meiosis, in repairing damaged duplex DNA, and in transferring characteristics such as antibiotic resistance between bacterial cells.

Key concepts in DNA recombination

- Generates new combinations of genes, or identical material can be exchanged, e.g. between sister chromatids
- Increases genetic variety on which natural selection can act—therefore very important in evolution

continued

DNA recombination

- Important in DNA repair when both strands of DNA damaged, as information for repair is derived from an undamaged duplex, e.g. sister chromatid
- Bacterial recombination exploited in molecular biology
- Frequent (e.g. one cross-over per 5 kb) in some yeast chromosomes; some hotspots of recombination in humans
- Major types of recombination:
    i. homologous recombination (HR)
    ii. non-homologous end joining (NHEJ)
    iii. transposition

General recombination

Process by which genetic cross-overs occur between two regions of DNA with considerable homology, via enzymes which show no sequence specificity. (Also known as homologous recombination.)

- Roles in:
    i. eukaryotic meiosis
    ii. repair of DNA double strand breaks
    iii. abrupt changes (switches) in genetic characteristics, e.g. in *S. cerevisiae* (brewers' yeast), *HO endonuclease* cleaves DNA, promoting recombination and allowing a ↔ α mating type switch
    iv. incorporation of novel DNA into bacterial chromosomes, e.g. after conjugation

(→) *see Homologous recombination (HR) (pp. 117–119) for mechanism (also Figure 4.22)*

Evidence for homologous recombination

- ssDNA intermediates in recombination, and *E. coli DNA pol I* and *DNA ligase* required for resolution of intermediates
- Electron microscopy of *E. coli* plasmids (cells treated with chloramphenicol to block chromosomal but not plasmid replication) shows 'figure of eight' structures. Cleavage of plasmids with unique site restriction enzyme Eco RI gave four-way junctions not two circles, suggesting four-way (χ) structures are intermediates in recombination
- *recA⁻* mutants do not form χ structures; therefore *recA* required in recombination in *E. coli*

Transposition

Transposition is the movement of a gene or transposable element (transposon) from one chromosome to a different location on the same chromosome, or to a different chromosome.

- Occurs in both prokaryotes and eukaryotes
- No extensive homology required for recombination (cf. general recombination requires long stretches of homology)
- **transposons** = mobile genetic elements
- Insertion of transposon into gene usually inactivates it, but may lead to transcription from adjacent gene via active transposase promoter
- Can result in insertions, deletions, knock-out or duplication of host genes
- Frequency ~10^{-6}/cell generation
- Requires specific enzyme—*transposase*, which cleaves DNA to form staggered cut at donor and recipient sites
- Simple insertion sequences (IS) of ~1 kb within the transposon encode *transposase* enzyme, i.e. the element encodes the enzyme needed for its own excision
- Terminal 20 residues of IS are same but in opposite orientation = inverted terminal repeats
- IS can move to any site on the chromosome—no need for homologous regions, and no role for homologous recombination proteins
- Requires *DNA polymerase I* and *DNA ligase* in *E. coli* in addition to *transposase*
- More complex transposons encode additional genes, e.g. Tn3 encodes β-*lactamase*, conferring resistance to penicillin-type antibiotics
- Medically important, e.g.
  - i. F factor in bacterial conjugation
  - ii. R factor for antibiotic resistance
  - iii. Salmonella flagellar antigens
- Used extensively in classical genetics, e.g. insertional mutagenesis experiments in *Drosophila melanogaster*
- Many naturally occurring transposons in eukaryotic genomes—usually silent

4.6 DNA EXCHANGE IN BACTERIA

Key concepts in DNA exchange in bacteria

- Horizontal transmission of genetic material usually between bacteria of same or related species
- Increases genetic diversity in bacteria—allows rapid adaptation to changes in environmental conditions, e.g. rapid spread of antibiotic resistance
- Three key types of exchange:
  - i. conjugation
  - ii. transduction
  - iii. transformation

Bacterial conjugation

- Also known as bacterial mating (Figure 4.23)
- 'Male' donor F$^+$ cell contains fertility F factor (plasmid), lacking in 'female' recipient F$^-$ cells
- Male and female cells associate via sex pili and then link via cytoplasmic conjugation tube
- F plasmid is nicked, then duplicated from the nick by rolling circle DNA replication
- ssDNA copy of F factor transferred through conjugation tube to 'female'
- Second DNA strand of F factor synthesized to reform duplex DNA in both cells—both now have F factor so are now 'male'
- Not all cells in a bacterial population become F$^+$ due to unequal partitioning of the plasmid on cell division (i.e. F factor inherited by only one daughter cell so now 2 cells, one F$^+$ the other F$^-$)

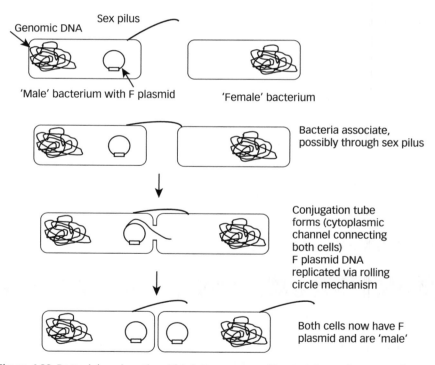

Figure 4.23 Bacterial conjugation. This is the process of bacterial sex where genetic information is transferred horizontally i.e. within the same generation. The F plasmid confers mating ability, but other plasmids can be transferred e.g. those encoding antibiotic resistance. If the F plasmid inserts into the bacterial genome, the bacterium is called Hfr (high frequency of recombination) and the entire genome can be transferred during conjugation. This is how the order of genes on the *E. coli* chromosome was originally discovered. Note that not all bacteria in a population become 'male' since the F plasmid is not partitioned to both daughter cells on cell division.

- If F factor integrates into host genome, Hfr (high frequency of recombination) strain produced—some or all of host chromosome adjacent to site of F factor integration is carried over into recipient cell, where extensive homologous recombination can occur between transferred DNA and recipient cell chromosome
- Very useful in coarse mapping of the order of genes on *E. coli* chromosome by assessing time taken for specific characteristics to enter recipient cell
- R factor similarly transferred, conferring resistance to antibiotics very rapidly in an entire bacterial population without a need for sexual mating to transmit the new genetic information—clinically important as can lead to rapid spread of antibiotic resistance

Evidence for bacterial conjugation
- Conjugation tube and sex pili visible by EM
- Can experimentally interrupt mating by agitating cells to break sex pili and conjugation tube—Waring bender experiments especially in Hfr strains; assess time taken for genes to be transferred by complementation of, for example, auxotrophy in F⁻ recipient cell

Transduction

- Transfer of DNA between bacteria via phage vectors
- Very useful in fine-mapping details of gene order in *E. coli* chromosome—frequency of two genes being carried together in same phage particle (co-transduction frequency) is inversely proportional to distance apart of those genes on *E. coli* chromosome
- Two types: generalized and specialized

1. Generalized transduction
- e.g. phage P1
- Phage-encoded *P1 nuclease* cuts rolling-circle replicated phage DNA into individual genome-sized chunks for packaging in phage heads
- *P1 nuclease* also cleaves host genomic DNA in lytic phase
- Host DNA fragments can become randomly packaged into phage heads—no sequence specificity (though packaged according to size)
- Large amount of bacterial DNA can be carried (~90 kb)

2. Specialized transduction
- e.g. phage lambda (λ)
- During establishment of lysogenic phase, phage inserts its own DNA in specific *att* site on host chromosome between *gal* and *bio* genes
- Phage DNA is excised from host chromosome in switch from lysogenic to lytic phase. Incorrect excision may result in uptake of some flanking host DNA (either

gal or *bio*) together with most of phage DNA, which is then packaged into phage particles that can infect other cells

- Only specific sequences can be transferred, i.e. those adjacent to the *att* site
- Only small regions of bacterial DNA can be carried (~5 kb)
- Phage lambda has proven a very useful vector in cDNA library formation and other cloning

Transformation

- Uptake of naked DNA by bacteria from their environment
- DNA from same or very closely related species is retained, while DNA from non-related species is cleaved by <u>restrict</u>ion enzymes (<u>restrict</u> host range for foreign DNA, e.g. bacteriophages)
  - ➔ *see Chapter 9: Restriction enzymes (p. 221)*
- Requires adhesion zones on bacterial surface; uptake is probably via membrane transporter
- Name derives from Griffith's early experiments on transforming non-pathogenic bacteria to pathogenic by incubation with heat-killed bacteria
  - ➔ *see Chapter 1, Figure 1.4, (p. 26)*
- Important in horizontal transmission of genetic characteristics (i.e. within same cell generation—no need for reproduction)
- Very widely used in molecular cloning to introduce foreign DNA fragments into *E. coli* cells for amplification, etc.
- Transformation frequencies can be greatly enhanced by chemical/thermal treatment of bacterial cells that aids association of negatively charged DNA with phospholipid membrane, then increases permeability of plasma membrane (e.g. treatment with rubidium chloride followed by brief heat shock)

Revision tip

Transformation has two meanings in biology so make sure you use it correctly. It can mean DNA uptake by <u>prokaryotes</u> as described above, but transformation of mammalian cells means that they have changed characteristics to become cancerous. Experimentally-induced uptake of DNA into <u>mammalian</u> cells is called **transfection**.

 Check your understanding:

How and why do bacteria take up DNA? (*Hint: remember to include conjugation, transduction, and transformation, with key example such as transfer of antibiotic resistance. You could also include how the DNA taken up can recombine with the host chromosome through homologous recombination; the 'why' part should include generating genetic diversity, acquiring favourable characteristics (e.g. antibiotic resistance, etc.)*

5 Mammalian metabolic pathways

5.1 ANABOLISM AND CATABOLISM

Key concepts in metabolism

- Catabolism = breakdown
- Anabolism = synthesis/building up
- ATP is energy currency of the cell; it is not stored
- Glucose is stored as glycogen; fats provide a very high density energy store
- Mammalian cells can synthesize/break down sugars, lipids, and amino acids (some amino acids cannot be synthesized and so are essential in the diet)
- Pyruvate and acetyl CoA (AcCoA) are major products of catabolism of carbohydrates, fats, and proteins:
    - i. under aerobic conditions, AcCoA feeds into **TCA cycle**
    - ii. resulting electrons from TCA cycle are passed down the electron transport chain and generate ATP through oxidative phosphorylation
- Major routes of ATP production in the cell are from:
    - i. glucose (6C) breakdown via glycolysis → TCA cycle → oxidative phosphorylation to yield ~30 ATP/glucose
    - ii. lipid breakdown via lipolysis; released fatty acids then broken down by β-oxidation, e.g. 16C palmitate yields ~108 ATP

continued

- FAD (reduced to $FADH_2$) and NAD^+ (reduced to NADH) are important **cofactors** in catabolism; NADPH used as reducing agent in biosynthesis
- Products of one metabolic pathway can be used as substrates in another pathway (see Figure 5.1)
- Amino acid breakdown via transamination yields glutamate + carbon skeletons that can be either glucogenic or ketogenic, plus ammonia
- Anaplerotic reactions 'fill up', i.e. replenish compounds used in other reactions
- Table 5.1 for comparison of catabolism and anabolism

	Catabolism	*Anabolism*
Energy	breakdown	biosynthesis
	ATP-generating	ATP-using
Cofactor	oxidant: NAD^+	reductant: NADPH
Glucose	glycolysis then TCA cycle and oxidative phosphorylation	gluconeogenesis
Glycogen	glycogenolysis then glycolysis, TCA cycle, and oxidative phosphorylation	glycogen synthesis
Lipids	lipolysis; β-oxidation, then TCA cycle, and oxidative phosphorylation	fatty acid synthesis, then triglyceride synthesis
Amino acids	transamination, then glucogenic or ketogenic pathways; ammonia detoxified via urea cycle	amino acid synthesis

Table 5.1 Key pathways in mammalian metabolism

ATP

- ATP is a good energy source:
  - i. *thermodynamically unstable* — the phosphate bonds have a high free energy of hydrolysis due to restriction in resonance states
  - ii. *kinetically stable* as phosphate bonds are stable in the absence of catalysis
- ATP also acts directly as a phosphate donor (e.g. formation of glucose 6-phosphate)
- ATP acts as an energy 'currency' in the cell—universal in biology
- ATP is *not* stored in the cell
- Different free energy from breaking one or two phosphate bonds in ATP:
  - $ATP + H_2O \rightarrow ADP + P_i$ $\Delta G^{0\prime}$ −30.5 kJ/mol
  - $ATP + H_2O \rightarrow AMP + PP_i$ $\Delta G^{0\prime}$ −45.6 kJ/mol
  - Hydrolysis of PP_i to $2P_i$ is also thermodynamically favourable (negative ΔG)

Catabolism

- **Catabolism** is the breakdown of organic molecules to release energy in a usable form of ATP, i.e. *catabolism is ATP-generating*

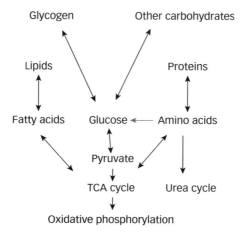

Figure 5.1 The interplay between the key components and pathways in metabolism. Note that pyruvate cannot be generated from the TCA cycle in animals.

- Carbon skeletons are broken down into one of a small number of basic units (e.g. pyruvate, acetyl CoA) that can be processed:
  - i. to generate more ATP (usually via TCA cycle), or
  - ii. used as building blocks for biosynthesis (in anabolism)
- Molecules are oxidized in stages to release energy in small amounts in reactions with negative ΔG
- Reactions with sufficient negative ΔG can then be *coupled* to the synthesis of ATP from ADP and orthophosphate (P_i)
- Organic molecules are often activated with the initial input of energy in order to make them reactive and facilitate **oxidation**
- Oxidation of carbon skeletons generates reducing power. Resulting electrons are passed to adenine nucleotide derivatives (FAD and NAD^+) as carriers
- The intermediate used depends on the value of reducing power generated: NADH + H^+ (generated by oxidation of alcohols and ketoacids) has a higher reduction potential than $FADH_2$ (generated by oxidation of $-CH_2-CH_2-$)
- NADH, and $-CH_2-CH_2-$ via FAD cofactors, pass electrons into the electron transport chain in the inner mitochondrial membrane to reduce O_2 to H_2O. Energy released in this stage generates the majority of ATP in the cell but can only take place in aerobic conditions when O_2 is available
- Different tissues use different energy sources:
  - i. glucose: red blood cells and brain
  - ii. fatty acids: heart

Anabolism

- **Anabolism** is the biosynthesis of complex molecules using energy, usually in form of ATP, i.e. anabolism is ATP-using

Anabolism and catabolism

- Irreversible hydrolysis of pyrophosphate ($PP_i \rightarrow 2P_i$) by **pyrophosphatase** can drive biosynthetic reactions in which PP_i (often from breakdown of ATP) is a product of an equilibrium reaction. Removing one of the products drives the equilibrium reaction in one direction, e.g. phosphodiester bond formation in DNA and RNA synthesis

  ➔ *see 4.1 DNA replication (p. 78) and 4.2 RNA synthesis (transcription) (p. 88) for more details*

- Biosynthetic pathways are NOT the reverse of the relevant catabolic pathways, even though they may share several steps
- NADPH is normally the reductant in biosynthesis (cf. NAD^+ is reduced by catabolic pathways)

Revision tip

- Remember: **OILRIG**: **O**xidation **I**s **L**oss, **R**eduction **I**s **G**ain (of electrons).

Metabolic regulation

- Distinct enzyme-catalysed pathways exist for synthesis and degradation
- Breakdown and synthesis of a molecule are coordinately regulated such that only one direction of a potential pathway is active in one compartment at any one time—prevents substrate cycling in most cases
- Shared steps are **equilibrium reactions**—direction of reaction is governed by relative concentrations of substrate and product
- **Irreversible reactions** are not shared and constitute major control points—usually where a molecule becomes committed to a particular fate
- Irreversible reactions are the controlled steps, allowing the cell to switch between breakdown and synthesis of a particular molecule in response to its needs
- Which pathway predominates is dependent on the energy charge in the cell:
    i. high: inhibit ATP-generating processes, i.e. catabolism is inhibited
    ii. low: inhibit ATP-using processes, i.e. anabolism is inhibited
- Three main mechanisms of regulating enzyme reactions:
    i. substrate availability
    ii. enzyme amount
    iii. enzyme catalytic activity (can be altered, for example, by phosphorylation)
- Enzyme activity can be regulated by:
    i. **allosteric** modulation
    ii. **covalent** modification (e.g. phosphorylation)
    iii. **compartmentalization** (keeping substrate away from enzyme)
- Enzymes respond to the local needs of the cell by allosterically interacting with potential products/substrates for the pathways—**feedback** and **feedforward regulation**
- Excess product will feed back to inhibit the enzyme catalysing the first committed step in its biosynthesis

- Interplay between different organs allows co-ordination of overall response of organism
  - ➔ *see 6.1 Key organs in metabolic integration (p. 168) for interorgan relationships in metabolism*
- The needs of the organism are signalled by hormones, e.g.
  - i. **insulin**: fed/high blood glucose
  - ii. **glucagon**: fasting/low blood glucose
  - iii. **adrenaline (epinephrine)**: muscle activity
- Hormones generally act by inducing pathways that change phosphorylation state of regulatory metabolic enzymes to stimulate or inhibit their activity
- Hormonal control often overrides allosteric control, e.g. muscle contraction: adrenaline stimulates glycogen breakdown in liver even though liver itself does not require extra glucose
  - ➔ *see 6.4 Key hormones in metabolic integration (p. 178) for hormone effects on metabolism*
- Long-term metabolic control can be exerted at level of synthesis of enzymes for metabolic pathways, especially in micro-organisms (e.g. lac operon)
  - ➔ *see 4.2 RNA synthesis (transcription) (p. 94) for lac operon*
  - and for gluconeogenic enzymes during starvation in mammalian cells

5.2 GLUCOSE BREAKDOWN AND SYNTHESIS

Glucose breakdown: glycolysis

Glycolysis is a pathway with ten enzyme-catalysed steps by which glucose (C6) is broken down to generate two molecules of pyruvate (C3) and two ATP per free glucose (though note that three ATPs are produced per glucose from glycogen). No oxygen is required. Glycolysis is a ubiquitous pathway in all cell types. Products from glycolysis can feed into the TCA cycle and thence oxidative phosphorylation to yield overall 30 molecules of ATP per glucose.

1. **How is glucose broken down?**
- Ten enzyme-catalysed steps which result in formation of pyruvate from glucose (Figure 5.2):
  - i. formation of glucose 6-phosphate (G6-P) by phosphate transfer from ATP catalysed by *hexokinase* (*glucokinase* in liver): IRREVERSIBLE STEP
    Phosphorylation important to:
    a. activate glucose
    b. accumulate G6-P in the cell when glucose levels outside are low (to maintain glucose concentration gradient for influx)
    c. prevent efflux of glucose (G6-P cannot cross plasma membrane)
  - ii. conversion of G6-P to fructose 6-phosphate (F6-P) by *phosphohexose isomerase* (also known as *glucose 6-phosphate isomerase*)

Glucose breakdown and synthesis

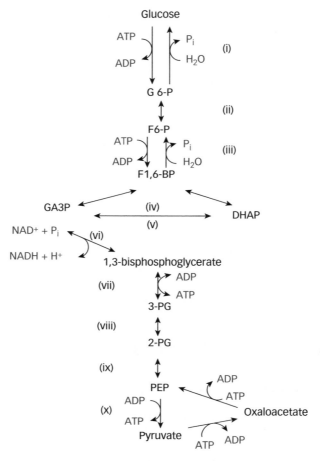

Figure 5.2 Glucose is converted into pyruvate via glycolysis and can be synthesized from pyruvate by gluconeogenesis. Equilibrium reactions are shared. Irreversible reactions are unique to synthesis or degradation. These are the reactions that use or generate energy, and which are controlled.

iii. phosphate from ATP added to F6-P to form fructose 1,6-bisphosphate (F1,6-BP) + ADP catalysed by *phosphofructokinase*: IRREVERSIBLE STEP

iv. F1,6-BP broken down by *aldolase* into 2 × 3 C compounds: glyceraldehyde 3-phosphate (GA3P) and dihydroxyacetonephosphate (DHAP)

v. DHAP converted to GA3P via *triose phosphate isomerase* (reaction to GA3P favoured as GA3P removed by subsequent glycolysis reactions, shifting equilibrium)

vi. GA3P converted to 1,3-bisphosphoglycerate by *glyceraldehyde 3-phosphate dehydrogenase* (reaction converts NAD$^+$ → NADH)

vii. ATP generated by donation of phosphate group from 1,3-bisphosphoglycerate to ADP, catalysed by *phosphoglyerate kinase*, forming 3-phosphoglycerate

    viii. conversion of 3-phosphoglycerate (3-PG) to 2-phosphoglycerate (2-PG) (near equilibrium) by *phosphoglycerate mutase*

    ix. dehydration of 2-phosphoglycerate to phosphoenolpyruvate (PEP) + H_2O by *enolase*

    x. substrate level phosphorylation to generate ATP from PEP + ADP, forming pyruvate, catalysed by *pyruvate kinase*: IRREVERSIBLE. NB *Pyruvate kinase* is tightly regulated allosterically, covalently, and at gene expression level by hormones

- Overall pathway generates two ATPs, two NADH, and two pyruvate per glucose
- Pyruvate has three possible fates depending on the degree of oxygenation and the species. It can be converted to:
  - i. acetyl CoA, which enters the TCA cycle under aerobic conditions; $2CO_2$ evolved
  - ➔ *see 5.3 Tricarboxylic acid (TCA) cycle (p. 137)*
  - ii. ethanol + $2CO_2$: anaerobic, e.g. some gut bacteria, brewers' yeast
  - iii. lactate: anaerobic, e.g. muscle (carried in blood and converted back to pyruvate in liver by hepatic *lactate dehydrogenase*) (see Figure 5.10)
- Essential that pyruvate is metabolized further in order to regenerate the NAD^+ required for further glycolysis to take place
- Pyruvate → acetyl CoA step:
  - i. irreversible in animals and the carbon skeleton cannot be used to remake glucose
  - ii. in plants and bacteria, this step can be reversed—the **glyoxylate cycle**
  - ➔ *see 8.2 Ancillary reactions of photosynthesis (p. 214) for the glyoxylate cycle*
- NADH generated in glycolysis can enter the electron transport chain (ETC)
  - ➔ *see 5.4 Oxidative phosphorylation (p. 140)*

2. Where is glucose broken down?

- Glycolysis is universal in biology
- Cytosol of all cell types:
  - i. red blood cells can use *only* glucose as fuel
  - ii. brain prefers glucose
  - iii. many tissues will use other fuels in preference, e.g. ketone bodies in heart
  - ➔ *see 5.7 Ketone body synthesis and breakdown (p. 154) for more information on ketone bodies*

3. When is glucose broken down? (see Table 5.2)

- When ATP is required for cell function (e.g. when demand rises in exercise), particularly in anaerobic tissues (red blood cells, vigorously contracting muscle)

4. Key evidence for glycolysis

- Cell-free yeast extract will convert sucrose to alcohol (Buchner and Buchner, 1897)

Glucose breakdown and synthesis

- In 1929, Harden isolated F1,6-BP from yeast extracts (Nobel prize for elucidating glycolysis); also showed requirement for P_i for glucose fermentation by yeast cell extracts
- Heat labile and heat stable components required—suggests involvement of enzymes and co-enzymes, respectively
- Full pathway elucidated 1940s—named Embden–Meyerhof pathway after two key researchers

Looking for extra marks?

Remember that other sugars can also enter glycolysis, e.g. fructose and galactose. Fructose is converted to F1-P, and then to DHAP and glyceraldehyde, both of which are then converted to GA3P and are metabolized in same way as GA3P arising from glucose.

Glucose synthesis: gluconeogenesis

Glucose levels in the blood must be maintained in order to provide fuel to brain and other organs. During prolonged starvation, when glycogen supplies are depleted, glucose must be synthesized from pyruvate or other glucogenic compounds (e.g. certain amino acids) in a series of reactions called gluconeogenesis.

1. **How is glucose synthesized?**
- Glucose can be synthesized from non-carbohydrate precursors by **gluconeogenesis,** usually starting from common substrate, pyruvate
- Equilibrium steps are shared with glycolysis (see Figure 5.2). The direction of the reaction is set by the relative levels of substrates and products
- Irreversible steps are <u>not</u> shared:

    i. pyruvate $+ ATP + HCO_3^- \rightarrow$ oxaloacetate $+ ADP + P_i$ (*pyruvate carboxylase*)

    ii. oxaloacetate $+ GTP \rightarrow PEP + GDP + CO_2$ (*PEP carboxykinase*) (note that reverse reaction of i and ii is catalysed by *pyruvate kinase*)

    iii. F1,6-BP $+ H_2O \rightarrow$ F6-P $+ P_i$ (*fructose 1,6-bisphosphatase*) (reverse reaction catalysed by *phosphofructokinase*)

    iv. G6-P $+ H_2O \rightarrow$ glucose $+ P_i$ (*glucose 6-phosphatase*) (reverse reaction catalysed by *hexokinase*)

- Six ATP equivalents are used to synthesize one molecule of glucose from pyruvate
- Pyruvate is derived from lactate (e.g. after intense exercise) and on breakdown of some amino acids (e.g. alanine)
- Oxaloacetate generated from breakdown of other glucogenic amino acids, e.g. tyrosine breakdown yields fumarate which is converted to oxaloacetate
- Oxaloacetate in the mitochondrial matrix is shuttled to the cytosol in the form of malate. Oxaloacetate is then regenerated in the cytosol for gluconeogenesis (reverse of malate shuttle—see Figure 5.3)

Enzyme	Regulator	Effect (+ve or −ve)	Mechanism	Signal	Notes
Hexokinase	glucose 6-phosphate	−	allosteric	product inhibition	glucose 6-phosphate can enter other pathways, e.g. glycogen synthesis, so this step does not commit glucose to glycolysis NB *glucokinase* = isoform of *hexokinase* found in liver; higher K_m and is not inhibited by product
Phosphofructokinase (PFK)	AMP	+	allosteric	energy charge is low, ATP required	first irreversible step—commitment to glycolysis, so major control point
	citrate	−	allosteric	biosynthetic precursors available	allows fatty acid oxidation to limit glucose utilization
	H⁺	+	allosteric	excess lactate (product)	in anaerobic conditions
	fructose 2,6-bisphosphate	+	allosteric	high blood sugar (insulin)	second messenger for hormones
Pyruvate kinase	fructose 1,6-bisphosphate	+	allosteric	substrate activation feed forward	controls outflow from glycolysis
	ATP	−	allosteric	energy charge	
	alanine	−	allosteric	biosynthetic precursors	
	protein kinase A (liver enzyme only)	−	phosphorylation	low blood sugar levels (glucagon)	
	protein phosphatase	+	dephosphorylation	high blood sugar (insulin)	

Table 5.2 Regulation of glycolysis

Chapter 5 Mammalian Metabolic Pathways 135

Glucose breakdown and synthesis

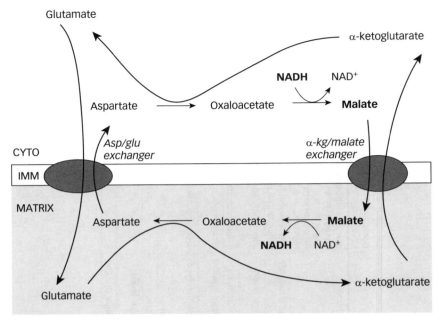

Figure 5.3 Malate shuttle. The role of the malate shuttle is to transport electrons [2H] across the inner mitochondrial membrane (IMM) into the matrix. Cytoplasmic oxaloacetate is converted to malate with oxidation of NADH to NAD$^+$. The malate passes through the IMM and is then oxidized back to oxaloacetate, generating NADH that can then enter the electron transport chain. Amino transferases reversibly convert the oxaloacetate and α-ketoglutarate (α-kg) to glutamate.

- Glycerol (e.g. from lipid breakdown) enters at DHAP (see Figure 5.2)
- Lactate and alanine formed by contracting muscle can be transported in blood and converted into glucose by the liver (**Cori cycle**—see Figure 5.18)

  ➔ *see 5.10 Lactate and ethanol metabolism (p. 164) for more on lactate metabolism*

2. Where is glucose synthesized?

- Only the liver and kidney can synthesize glucose in mammals ('altruistic organs')—they export it via blood to glucose-requiring tissues, e.g. brain and contracting muscle
- Most reactions are cytosolic
- *Pyruvate carboxylase* is mitochondrial

3. When is glucose synthesized?

- Fasting/starvation (for regulation—see Table 5.3)
- Periods of intense exercise
- i.e. when blood glucose is low and cannot be replenished by glycogen breakdown

Enzyme	Regulator	Effect (+ve or –ve)	Mechanism	Signal	Notes
Fructose 1,6-bisphos-phatase	AMP	–	allosteric	energy charge	catalyses reaction opposing *PFK* so reciprocal regulation (Figure 5.2)
	citrate	+	allosteric	biosynthetic precursors	
	fructose 2,6-bis-phosphate	–	allosteric	excess substrate for glycolysis	
Pyruvate carboxylase (PC)	ADP	–	allosteric	energy charge is low, glycolysis should predominate over gluconeogenesis	reciprocal control to *PFK* (with *PEPC*)
	acetyl CoA	+	allosteric	biosynthetic precursors	
Phospho-enolpyruvate carboxykinase (PEPC)	ADP	–	allosteric	energy charge	reciprocal control to *PFK* (with *PC*)

Table 5.3 Regulation of gluconeogenesis

Looking for extra marks?

Glucose 6-phosphatase is inside the endoplasmic reticulum; its action releases glucose into the blood.

Check your understanding

Outline the differences between the pathways used for the synthesis of glucose from pyruvate, and its breakdown to pyruvate. (*Hint: show metabolic pathways as diagrams and highlight irreversible steps and non-shared steps, e.g. by using a different colour. Remember to include regulation.*)

5.3 TRICARBOXYLIC ACID (TCA) CYCLE

The TCA cycle is responsible for the complete oxidation of acetyl CoA to CO_2.

Key concepts of the TCA cycle

- Also known as the **Krebs cycle** or the **citric acid cycle**
- Occurs in the mitochondria in eukaryotes
- Acetyl CoA (2C) is initially attached to a 4C acid (oxaloacetate) to form citrate (6C—the 'tricarboxylic acid')
- Two carbon atoms are sequentially removed by oxidative decarboxylation

continued

Tricarboxylic acid (TCA) cycle

- Oxaloacetate is regenerated
- Several of the intermediates of the TCA cycle are also used/generated in other pathways, e.g. gluconeogenesis, amino acid degradation, amino acid synthesis, urea cycle, etc.

Supply of acetyl CoA

- Before pyruvate (e.g. from glucose, lactate) can be oxidized in the TCA cycle, it is oxidatively decarboxylated to acetyl CoA (producing NADH + CO_2) in the **link reaction** by *pyruvate dehydrogenase*
- Acetyl CoA is also produced directly as a result of fatty acid oxidation and breakdown of aliphatic amino acid side chains
  - ➔ *see 5.6 Lipid breakdown (p. 149) for β-oxidation pathway*
- Acetyl CoA enters the TCA cycle by condensation with oxaloacetate to form citrate

Reactions of the TCA cycle

- See Figure 5.4 for TCA cycle
- Two oxidative decarboxylations (*isocitrate dehydrogenase* and *2-oxoglutarate dehydrogenase*) each generate NADH and release CO_2
- CO_2 release renders these reactions irreversible—important for regulating the cycle
- Cycle generates reducing power (NADH and $FADH_2$), which is the major source of ATP production via the electron transport chain (ETC)
  - ➔ *see 5.4 Oxidative phosphorylation (p. 140) for oxidative phosphorylation and the ETC*
- Three molecules of NADH and one of $FADH_2$ are generated per acetyl CoA
- Oxidation of these cofactors by O_2 generates approximately nine ATP per acetyl CoA
- One high energy phosphate bond (as GTP) is formed directly in the *succinyl thiokinase* reaction
- GTP can be used to generate ATP in the equilibrium reaction: ADP + GTP ↔ ATP + GDP
- The TCA cycle is a source of many carbon skeletons for biosynthesis (Figure 5.4)
- For regulation of TCA cycle see Table 5.4

Key evidence for TCA cycle

- Addition of an intermediate (e.g. malate) to a cell-free extract oxidizing pyruvate gives a very large increase in CO_2 production, much larger than the amount of malate added
- This 'catalytic' effect of malate on pyruvate oxidation shows that malate is not consumed but must be regenerated in a 'cycle'

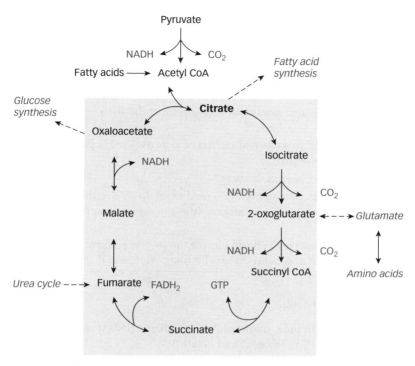

Figure 5.4 Tricarboxylic acid cycle (shaded) showing links to other major pathways.

Enzyme	Regulator	Effect (+ ve or –ve)	Mechanism	Signal	Notes
Pyruvate dehydrogenase	pyruvate dehydrogenase kinase	–	phosphor-ylation	high energy charge, reducing power	controls the production of acetyl CoA
	pyruvate dehydrogenase phosphatase	+	dephos-phorylation	hormonal	e.g. adrenaline in heart tissue
Pyruvate dehydrogenase kinase	high NADH/NAD$^+$ ratio	+	allosteric	excess reducing power	
	high acetyl CoA/ CoA ratio	+	allosteric	prevents glucose utilization during FA oxidation	
Pyruvate dehydrogenase phosphatase	Ca^{2+}	+	allosteric	hormonal, e.g. vasopressin	muscle contraction raises Ca^{2+}
Isocitrate dehydrogenase	Ca^{2+}	+	allosteric	intrinsic Ca^{2+} release from internal stores	may involve hormone action
	ADP	+	allosteric	energy charge	
α-Ketoglutarate dehydrogenase	Ca^{2+}	+	allosteric	intrinsic Ca^{2+} release from internal stores	muscle contraction raises Ca^{2+}
	high NADH/NAD$^+$ ratio	–	product inhibition	low respiration rate	low ADP inhibits respiration

Table 5.4 Regulation of the TCA cycle

5.4 OXIDATIVE PHOSPHORYLATION

Oxidative phosphorylation is the major pathway of ATP production in aerobic cells.

> ## Key concepts of oxidative phosphorylation
>
> - Electrons are generated as reduced cofactors from glycolysis and the TCA cycle
> - They are passed down a chain of carriers of increasing redox potential (the electron transport chain)
> - Finally they reduce oxygen to water
> - Energy released in electron transfer is used to set up an electrochemical gradient of protons across a membrane (the inner mitochondrial membrane in eukaryotes)
> - Special channels couple the dissipation of this proton gradient to the generation of ATP, and transport of ATP (out), ADP, and P_i (inwards)

Electron transport chain

- Protein complexes in inner mitochondrial membrane (IMM) comprise electron transport chain (ETC) (Table 5.5 and Figure 5.5)
- Three of these serve as electron carriers between NADH and O_2:
    i. complex I (*NADH dehydrogenase*)
    ii. complex III (cytochrome bc_1)
    iii. complex IV (*cytochrome oxidase*)
- Electron transfer complexes contain iron–sulphur clusters, flavins, **haem**, and/or copper ions. Electron carriers containing haem are called **cytochromes**
- NADH (a diffusible cofactor) or $FADH_2$ (an enzyme-bound cofactor) are generated in TCA cycle in the mitochondrial matrix by *dehydrogenases*
- Electrons are transferred from these cofactors to O_2 by the ETC electron carrier complexes linked by two mobile electron carriers—ubiquinone (Q) and cytochrome c
    i. QH_2 passes on its electrons one at a time to link two-electron (organic) carriers to one-electron (metal ion) carriers
    ii. cytochrome b is essential for splitting the two electrons
- Electrons e^- from other substrates (succinate, fatty acids) enter the chain via other flavoprotein complexes (e.g. *succinate dehydrogenase*–'complex II', *FA-CoA dehydrogenase*). All these electrons converge at Q, and their path is common from this point on (see Figure 5.5)
- Protons are pumped outwards from the mitochondrial matrix towards the cytoplasm
- **Proton motive force** (pmf) generated as a gradient of protons across inner mitochondrial membrane (IMM)—the matrix has an electrical potential negative (N side) relative to the outside (P side)

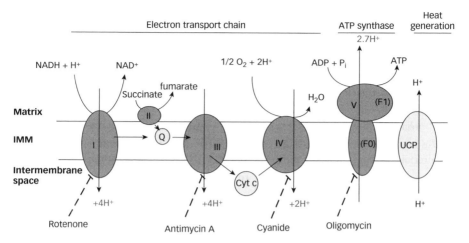

Figure 5.5 Organization of the oxidative phosphorylation system in the mitochondrial membrane. This is a one-dimensional representation; in reality, the complexes are able to move in the fluid bilayer in two dimensions. Note that H⁺ is lost from the matrix and gained by the intermembrane space. In brown adipose tissue, an uncoupling protein (UCP, light grey) spans the inner mitochondrial membrane and allows H⁺ to flow down their concentration gradient, generating heat.

	Name	Proton pump?	H⁺ pumped per 2e⁻	
Complex I	NADH dehydrogenase (aka NADH-ubiquinone oxidoreductase)	yes	4 H⁺	NADH
Complex II	succinate dehydrogenase	no	None	succinate via FADH₂
Complex III	cytochrome bc (aka ubiquinol-cytochrome c oxidoreductase)	yes	4 H⁺	
Complex IV	cytochrome c oxidase	yes	2 H⁺	
Complex V	ATP synthase	yes	generates ATP	
				Overall: 10 H⁺ per NADH

Table 5.5 Complexes of the electron transport chain

- The number of H⁺ ions (protons) pumped across the membrane per 2e⁻ passing through a complex is controversial but can be roughly calculated (see Table 5.5)
- Fewer H⁺ in total are pumped by succinate oxidation than by NADH oxidation, generating fewer molecules of ATP per 2e⁻

Looking for extra marks?

Note the remarkable similarity of ETC in oxidative phosphorylation to the electron transport chain of chloroplasts in photosynthesis.

➔ *see 8.1 Photosynthesis (p. 203) for light dependent reactions of photosynthesis*

Oxidative phosphorylation

ATP synthesis

- *ATP synthase* (complex V) is made up of channel (F_0) and catalytic (F_1) components (see Figure 5.5)
- Energy in proton motive force (pmf) used for ATP synthesis (by F_1) when the protons flow back across the membrane down their electrochemical gradient (through F_0)
- Proton flow leads to the release of tightly bound ATP from the enzyme
- Rotary catalysis: three positions of *ATP synthase* stalk lead to three active sites alternating through loose, tight, and open (L, T, O) conformations
- Number of H^+ flowing for full 360° turn ($\rightarrow$ 3ATP) varies between organisms, depending on the composition of the F_0 channel
    - i.   probably $8H^+$ in mammals
    - ii. higher in bacteria
- ADP (and P_i) transported into the mitochondria; ATP exported by *exchange translocases*
- ADP has three negative charges but ATP has four, so this exchange *uses* energy from the electrochemical gradient (effectively $1H^+$/ATP)
- Hence in mitochondria, 3.7 H^+ ((8/3) + 1) flow back down gradient to generate one ATP; predicted values are:
    - i.   ~2.7 ATP per NADH (10 H^+ pumped by ETC)
    - ii. ~1.7 ATP per $FADH_2$ (6 H^+ pumped by ETC)
- Slightly less ATP is generated than predicted from the proton stoichiometry of the *ATP synthase*— due to various losses including:
    - i.   NADH shuttle into the mitochondrion. NADH generated in the cytosol (e.g. in glycolysis) must be shuttled into the mitochondria as malate using energy (malate/aspartate shuttle—see Figure 5.3)
    - ii. **uncoupling proteins**, e.g. in heat generation in brown adipose tissue of hibernating animals and infant mammals (these act essentially to 'short-circuit' proton flow across the IMM). These proteins span the inner mitochondrial membrane and allow protons to 'leak' down their concentration gradient, generating heat

Regulation of the ETC

- ETC is controlled by availability of substrates:
    - i.   reducing power in form of NADH (controlled via control of TCA cycle enzymes)
    - ii. ADP—electrons do not flow unless ADP is simultaneously phosphorylated to ATP ('respiratory control').
    - iii. oxygen—electrons will not flow if O_2 not present as it is the final electron acceptor

Key evidence for oxidative phosphorylation and the ETC

- Isolated mitochondria can use oxygen to generate ATP
- Imposing a pH gradient across the inner mitochondrial membrane will drive ATP synthesis in the absence of electron transport ('acid bath' experiment)
- Electron transport in the absence of ATP synthesis generates a measurable proton gradient
- Rotation of *ATP synthase* stalk can be visualized by attaching fluorescent actin filament to F_1 immobilized (inverted) on a glass slide—can see rotation by fluorescence microscopy (John Walker and Paul Boyer, Nobel prize 1997)
- Specific inhibitors block the electron transport chain at different points (measure decrease in oxygen consumption rate) allowing the order of components to be determined (see Table 5.6)

Inhibitor	Action	Notes
Rotenone	Inhibits electron transfer through complex I	Blocks the chain if NADH is the source of electrons, but not if succinate is used
Antimycin A	Inhibits electron transfer through complex III	Bypassed by reducing cyt c directly with ascorbate
Cyanide	Inhibits electron transfer through complex IV	Blocks ETC irrespective of source of electrons; hence use as poison
2,4-dinitrophenol (DNP) (uncoupler)	Collapses proton gradient, and inhibits ATP synthesis	Electron transfer speeds up. Energy wasted as heat
Oligomycin	Blocks H^+ channel and inhibits ATP synthesis	Electron transfer inhibited. ET restored by uncoupler, e.g. DNP

Table 5.6 Inhibitors of the electron transport chain

Looking for extra marks?

Cancer cells shift metabolism away from oxidative phosphorylation towards glycolysis—the 'Warburg effect'. Many cancer cells in a solid tumour exist in a hypoxic environment and so use anaerobic metabolism.

 Check your understanding

How is ATP synthesized in mitochondria? (*Hint: remember both the TCA cycle and oxidative phosphorylation: you should also be able to discuss β-oxidation of fats.*)

5.5 GLYCOGEN BREAKDOWN AND SYNTHESIS

Glycogen is a branched chain polymer of glucose which constitutes a fuel store in both liver and muscle. It is synthesized as an energy store in both liver and muscle, when glucose is abundant. Under hormonal control, glycogen can be broken down to produce glucose when required via glycogenolysis.

Glycogen breakdown and synthesis

Glycogen breakdown: glycogenolysis

1. **How is glycogen broken down?**
- Glycogen is highly branched to allow rapid breakdown from many 4′ OH ends
- Glycogen α-1,4 glycosidic bonds (~90% of bonds in glycogen) are broken by *glycogen phosphorylase*, yielding glucose 1-phosphate
    - i. enzyme catalyses removal of glycosyl residues from non-reducing end of glycogen (see Figure 5.6)
    - ii. bond cleavage by addition of phosphate (P_i)—termed **phosphorolysis**
- *Phosphorylase* cannot cleave α-1,6 glycosidic bonds (~10% bonds in glycogen):
    - i. *transferase* moves three glycosyl residues from a branch point to the terminus, exposing an α-1,6 link at the original branch point
    - ii. this is hydrolysed by *α-1,6-glucosidase* (also called *debranching enzyme)* to generate free glucose
    - iii. glucose is then phosphorylated by *hexokinase*
- NB In eukaryotes, *transferase* and *α-1,6-glucosidase* present in same polypeptide—a bifunctional enzyme
- Glucose 1-phosphate is converted to glucose 6-phosphate by *phosphoglucomutase*
    - i. can enter glycolysis directly

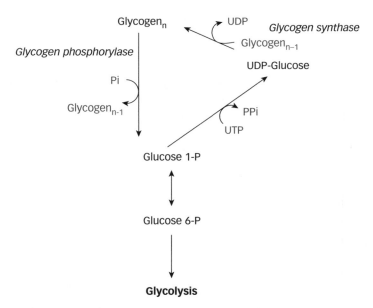

Figure 5.6 Glycogen metabolism. Glycogen is synthesized from, and degraded to, glucose 1-phosphate by different pathways. Glucose 1-phosphate can be converted to glucose 6-phosphate to enter glycolysis.

    ii. can be used to regenerate glucose from glucose 6-phosphate in liver via *glucose 6-phosphatase* and exported from liver cells to maintain blood glucose levels

- Glycogenolysis is regulated by phosphorylation of enzymes (see Table 5.7)

2. Where is glycogen broken down?

- In the cytoplasm of liver, muscle, brain, and kidney cells

3. When is glycogen broken down?

- When blood glucose levels drop, e.g. on fasting or starvation (liver)
- When demand for glucose is very high (exercising muscle)
- Signalled by high levels of the hormones glucagon (fasting) or adrenaline (exercise)
- Adrenaline (in muscles) or glucagon (in liver) bind membrane receptor
  - i. G protein activation on hormone binding of receptor activates *adenylate cyclase*
  - ii. *adenylate cyclase* converts ATP to cAMP which activates *protein kinase A* (PKA)
  - iii. *PKA* phosphorylates and activates *phosphorylase kinase*
  - iv. *phosphorylase kinase* phosphorylates and activates *phosphorylase* from inactive 'b' form to active 'a' form—can now break down glycogen

  ➔ *see 6.4 Key hormones in metabolic integration (p. 178) for hormonal regulation of glycogen metabolism*

Glycogen synthesis: glycogenesis

1. How is glycogen synthesized?

- Glycogen synthesis is not simply a reversal of the breakdown pathway—uses UDP-glucose as activated glucose donor (see Figure 5.6)
- Glucose 6-phosphate is converted to glucose 1-phosphate
- Glucose 1-phosphate is activated by linking to the nucleotide UTP to form UDP-glucose + PP_i catalysed by *UDP-glucose pyrophosphorylase*. Essentially IRREVERSIBLE, because the reaction (G1-P + UTP → UDP-G + PP_i) is driven to completion by hydrolysis of PP_i
- *Glycogen synthase* catalyses the addition of UDP-glucose residues via α-1,4 glycosidic bonds to:
  - i. **glycogenin** polypeptide primer—a dimeric enzyme where each monomer adds eight glucose units to its dimeric partner (glycogen synthase cannot add to glycogen chains of <4 glucose residues), or
  - ii. a pre-existing glycogen particle, if >4 residues long
- UDP is released
- A *branching enzyme* transfers a block of polymerized glucose units, forming α-1,6 glycosidic bonds at branch points—important to:
  - i. increase glycogen solubility

Glycogen breakdown and synthesis

    ii. generate large number terminal residues available for rapid synthesis and/or breakdown of glycogen

2. Where is glycogen synthesized?

- In the cytoplasm of liver, muscle, brain, and kidney cells

3. When is glycogen synthesized?

- In the fed state, i.e. when blood glucose levels are high
- Signalled by high levels of the hormone insulin—promotes glucose uptake into muscle cells, and activity of *glycogen synthase* in all cells
- Glycogen synthesis is <u>inhibited</u> by adrenaline and glucagon—phosphorylation cascade <u>inactivates</u> *glycogen synthase* (i.e. opposite from *glycogen phosphorylase* in breakdown)

Enzyme	Regulator	Effect (+ve or –ve)	Mechanism	Signal
Glycogen synthase	Protein kinase A (PKA)	–	phosphorylation	Fasting/starvation or exercise (glucagon or adrenaline, respectively)
	Protein phosphatase 1 (PP1)	+	dephosphorylation	High blood glucose (insulin)
Glycogen phosphorylase	Phosphorylase kinase	+	phosphorylation	Fasting/starvation (liver—glucagon) or exercise (liver and muscle—adrenaline)
	AMP	+	allosteric	Low energy charge, so need fuel for glycolysis to generate energy.
	Glucose 6-phosphate	–	allosteric	Glucose available so none required from glycogen breakdown
Phosphorylase kinase	PKA	+	phosphorylation	Fasting/starvation (liver—glucagon) or exercise (liver and muscle—adrenaline)
	PP1	–	dephosphorylation	High blood glucose (insulin)
	Ca^{2+}	+	allosteric	Muscle contraction releases Ca^{2+} into cytoplasm—glucose required for glycolysis etc. to provide energy

Table 5.7 Regulation of glycogen metabolism

Looking for extra marks?

Overall efficiency of energy storage in glycogen is ~97% since nine out of ten residues are cleaved to G1-P then converted to G6-P at no energetic cost. Free glucose released from branch points must be phosphorylated. Storage uses just over 1 ATP per G6-P stored; complete oxidation of G6-P gives ~31 ATP.

Glycogen storage diseases from mutation of *phosphorylase* results in intolerance to exercise; mutation of branching enzyme is usually fatal in infancy.

Check your understanding

Which reactions are the key control points of glycogen synthesis and breakdown? How are they regulated? (*Hint: remember control by the needs of the organism (hormonal control) as well as by the cell, e.g. energy requirement.*)

5.6 LIPID BREAKDOWN AND SYNTHESIS

Lipid breakdown

Lipids, in the form of triacylglycerols (TAG), are the major long term metabolic energy store in mammals. Triacylglycerols are derived from the diet or released from storage in adipocytes. Lipolysis breaks down fats to fatty acids and glycerol. Fatty acids are then broken down in 2C units via the β-oxidation pathway. Fats are highly reduced and can yield about 2.5 times more ATP per gram than glucose (higher 'energy density').

1a. How are dietary lipids broken down?
- Triacylglycerols (TAG) in diet must be broken down for transport across epithelial membrane by *lipases*:
    - i. hydrolysis (**lipolysis**) at C3, then C1 positions by *pancreatic lipases* in gut releases two fatty acid chains leaving 2-monoacylglycerol which can cross intestinal membrane
    - ii. TAGs resynthesized in intestinal mucosal cells
    - iii. packaged into **lipoprotein** particles (**chylomicrons**) and released into lymph and blood
- Membrane-bound *lipases* (*lipoprotein lipase*) on adipose cells cleave TAGs to free fatty acids and monoacylglycerols for transport into adipose tissue (regenerated and stored as TAGs)
- Lipolysis by cytoplasmic adipose *lipases* releases free fatty acids + glycerol (Figure 5.7)
- Free fatty acids (FFA) released into blood for uptake by energy-requiring tissues (muscle, heart)
- FFA not soluble in blood so bind to serum albumin in blood for transport
- Glycerol taken up by liver and converted to glycerol 3-phosphate then to DHAP—can be used in either glycolysis or gluconeogenesis

1b. How are fatty acids broken down?
- Three stages of fatty acid oxidation in vertebrates:
    - i. activation of fatty acids in cytosol (fatty acid + CoA + ATP → acyl CoA + AMP + PP$_i$)
    - ii. transport into mitochondria (carnitine + acyl CoA → acyl carnitine + CoA)
    - iii. degradation as two carbon units in mitochondrial matrix (β-**oxidation** pathway): acyl CoA (C_n) → acyl CoA (C_{n-2}) + acetyl CoA

Lipid breakdown and synthesis

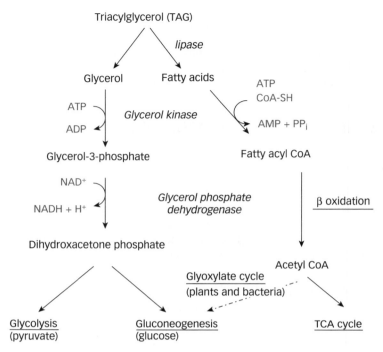

Figure 5.7 Overview of triacylglycerol breakdown. Triacylglycerols (TAGs) are initially broken down into glycerol and fatty acids. 2C units are successively removed from the fatty acids by β-oxidation to generate acetyl CoA, which can enter the TCA cycle. The glycerol is converted into dihydroxyacetone phosphate which can enter glycolysis/gluconeogenesis.

Activation of FA

- Two-step reaction catalysed by *acyl CoA synthetase*, breaks two phosphoanhydride bonds and therefore consumes 2 ATP equivalents:
    - i. fatty acid + ATP → acyl adenylate + PP$_i$ (IRREVERSIBLE because of PP$_i$ hydrolysis)
    - ii. acyl adenylate + HS-CoA → acyl CoA + AMP

Transport of fatty acyl CoA into mitochondria

- Requires transport shuttle—regulatory step (see Figure 5.8)
- Fatty acyl CoA is converted to acyl carnitine by *carnitine palmitoyl transferase I* (*CPTI*) on outer mitochondrial membrane
- Acyl carnitine shuttled across IMM by *translocase*
- *Carnitine palmitoyl transferase II* (*CPTII*) on matrix face of inner mitochondrial membrane transfers back acyl group to recreate fatty acyl CoA in mitochondrial matrix

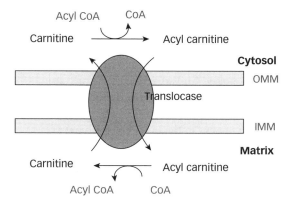

Figure 5.8 Transport of acyl CoA into the mitochondria. Transport involves two *carnitine acyl transferases*: I in the outer membrane and II in the inner membrane.

β-oxidation

- Fatty acyl CoA is broken down by repeated cycles of β-oxidation pathway to acetyl CoA, shortening the acyl CoA by two carbons each cycle, through oxidation, hydration, oxidation and thiolysis (see Figure 5.9)
    - i. first oxidation of acyl CoA (C_n) → trans Δ^2-enoyl CoA by *acyl CoA dehydrogenase*, reduces FAD to $FADH_2$ (donated to ETC for synthesis ~1.5 ATP)

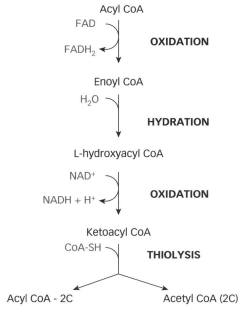

Figure 5.9 Overview of β oxidation. Acyl CoA is broken down by successive removal of two carbon units in the form of acetyl CoA. This series of four reactions is repeated until two acetyl CoAs are produced from even chain fatty acids, or one acetyl CoA and one propionyl CoA (three carbons) for odd chain fatty acids.

    ii. hydration of trans Δ^2-enoyl CoA $\rightarrow$ L-3-hydroxyacyl CoA by *enoyl CoA hydratase* (stereospecific forming only L isomer of 3-hydroxyacyl CoA)

   iii. second oxidation of L-3-hydroxyacyl CoA to 3-ketoacyl CoA by *L-3-hydroxyacyl CoA dehydrogenase*—reduces NAD^+ to NADH

   iv. thiolysis of 3-ketoacyl CoA to acyl CoA (C_{n-2}) + acetyl CoA by *thiolase*

- Remaining fatty acid (in form of acyl CoA) is shorter by two carbons: re-enters oxidation pathway
- Enzymes involved are present in the mitochondrial matrix
- Generates large amounts of energy
  - i. $FADH_2$ and NADH contribute electrons to ETC
  - ii. acetyl CoA can enter TCA cycle (if fat and carbohydrate degradation are balanced such that sufficient oxaloacetate molecules available for condensation with acetyl CoA)
- Animals cannot use acetyl CoA from fatty acid breakdown in gluconeogenesis
- Plants and many bacteria can use acetyl CoA for gluconeogenesis due to **glyoxylate cycle**

  $\circledast$ *see 8.2 Ancillary reactions of photosynthesis (p. 215) for the glyoxylate cycle*

β-oxidation of unsaturated FA
- Breakdown is same as saturated FA until reach double bond then:
  - i. *isomerase* converts *cis* enoyl CoA to *trans* double bond
  - ii. *trans* enoyl CoA is then substrate in oxidation pathway
- Polyunsaturated FA: hydration by *enoyl CoA hydratase*, then isomerization by *epimerase*

β-oxidation of odd numbered FA
- Oxidized by same β-oxidation pathway as even chain FA, except that final products = propionyl CoA + acetyl CoA
- Mammalian liver: propionyl CoA converted to succinyl CoA (see Figure 5.10) using biotin as cofactor

2. Where are lipids broken down?
- Dietary triacylglycerols (TAG) are hydrolysed to free FA and monoacyl glycerol in the gut
- TAGs are broken down to free FA and glycerol in adipose tissue, and released to blood
- Different compartments for each of the key reactions:
  - i. cytosol: activation of fatty acids
  - ii. mitochondrial matrix for β-oxidation

3. When are lipids broken down?
- Promoted by high levels of the hormone glucagon on fasting or starvation (see Table 5.8)
- Inhibited in fed state (via insulin, which promotes re-esterification)

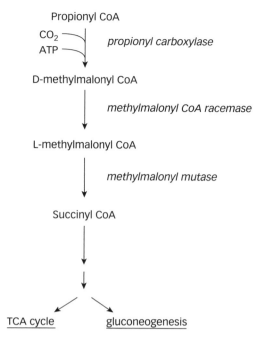

Figure 5.10 Conversion of propionyl CoA. Propionyl CoA generated from fatty acids with an odd number of carbons can be converted to succinyl CoA and enter either the TCA cycle or gluconeogenesis.

- FA are used as an oxidative fuel preferentially to glucose, particularly in starvation
- FA oxidation in muscle inhibits glucose oxidation

4. **Key evidence for β oxidation of FA**
- Straight chain fatty acids, with ω-carbon artificially linked to phenyl group, fed to dogs (Knoop)
- Analysis of derivatives in urine revealed phenylacetic acids when fed even-number C chain FA and benzoic acids when fed odd-number C chain FA

5. **Key example: oxidation of palmitate (16:0)**
- Palmitoyl CoA + 7FAD + 7NAD$^+$ + 7CoA + 7H$_2$O → 8 acetyl CoA + 7FADH$_2$ + 7NADH + 7H$^+$
- ATP yield 106 (see Table 5.9)

Looking for extra marks?

Defect in *translocase* (the enzyme that shuttles acyl carnitine across the IMM into the mitochondrial matrix) leads to muscle cramps on exercise, fasting or high fat diet.

Infants depend to a large extent on fat metabolism (major fuel in milk).

Lipid breakdown and synthesis

Enzyme	Regulator	Effect (+ve or −ve)	Mechanism	Signal	Notes
AcetylCoA carboxylase	citrate	+	allosteric	biosynthetic precursors	committed step in FA synthesis
	palmitoyl CoA	−	allosteric	excess FA	
	AMP	−	allosteric	low energy charge	FA breakdown should predominate over synthesis
	PP2A	+	dephosphorylation	insulin	
	PKA	−	phosphorylation	fasting/starvation (glucagon) or exercise (adrenaline)	
Carnitine acyl transferase I (CPT1)	malonyl CoA	−	allosteric	ongoing FA synthesis inhibits transfer to site of breakdown	CPT1 transports FA into mitochondria for β oxidation
Triacylglycerol lipase (adipose tissue only)	PKA	+	phosphorylation	requirement for energy—fasting/starvation (glucagon) or exercise (adrenaline)	FA released into bloodstream for use by other tissues
	PP1	−	dephosphorylation	plenty of available energy, so can store fats (Insulin)	

Table 5.8 Regulation of fatty acid metabolism

Component	ATP yield	Process	Total from FA
FADH$_2$	~1.5	ETC at ubiquinol	10.5
NADH	~2.5	ETC at complex I	17.5
Acetyl CoA	10	TCA cycle	80
Generation of palmitoyl CoA	−2	TCA cycle	−2
Overall yield from FA breakdown			106

Table 5.9 ATP yield from breakdown of palmitate

Fatty acid synthesis

Fatty acids are synthesized when resources are abundant and stored as triacylglycerides in adipose tissue.

1. **How are fatty acids synthesized?**
- Acetyl CoA = precursor for fatty acid synthesis
- IRREVERSIBLE committed step is conversion of acetyl CoA + ATP + HCO$_3^-$ → malonyl CoA + ADP + P$_i$ by *acetyl CoA carboxylase* and requires biotin (vitamin B$_7$)
- Elongation cycle (Figure 5.11):
  - i. acetyl CoA + acyl carrier protein (ACP) converted to acetyl-ACP by *acetyl transacylase*

    ii. malonyl CoA + acyl carrier protein (ACP) converted to malonyl-ACP by *malonyl transacylase*

    iii. condensation reaction: acetyl-ACP (2C) + malonyl-ACP (3C) → acetoacetyl-ACP (4 C) + ACP + CO_2 (see Figure 5.11)

- *Fatty acid synthase* is a multifunctional enzyme of two polypeptide chains in higher organisms
  - ➔ *see 3.6 Multienzyme complexes (p. 74)*
    - i. acyl carrier protein (ACP) carries substrates from one active site to the next on *fatty acid synthase*—permits co-ordination of multiple enzyme activities
    - ii. intermediates in FA synthesis are linked to ACP via sulphydryl groups

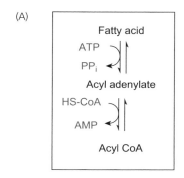

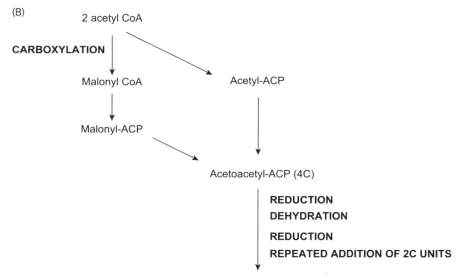

Figure 5.11 Synthesis of fatty acids. (A) Activation of fatty acids. This occurs on the outer mitochondrial membrane in a two-step process that consumes one molecule of ATP. (B) Malonyl ACP and Acetyl ACP are used to generate a 4C unit of acetoacetyl ACP. Repeated addition of 2C units is used to build up long chain fatty acids.

- Fatty acid synthesis requires NADPH for reductions—generated from pentose phosphate pathway and/or *malic enzyme*
  - i. one NADPH from one NADH by *malic enzyme*
  - ii. two NADPH/glucose in **pentose phosphate pathway**
- *Fatty acid synthase* carries out repeated addition of 2C units in cycle of reactions:
  - i. reduction of acetoacetyl-ACP by NADPH to hydroxybutyryl-ACP + $NADP^+$
  - ii. dehydration of hydroxybutyryl-ACP to crotonyl-ACP + H_2O
  - iii. reduction of crotonyl-ACP by NADPH to butyryl-ACP + $NADP^+$
- Elongation until C16-acyl-ACP formed—substrate for *thioesterase* to form palmitate and ACP

2. When are fatty acids synthesized?
- When carbohydrates or amino acids are plentiful (see Table 5.8)
- In dietary excess of glucose, fats can be made from glucose (but note can't make glucose from fats in animals)

3. Where are fatty acids synthesized?
- Acetyl CoA, precursor for fatty acid synthesis, is formed from pyruvate in mitochondria, especially in liver and adipose tissue, and converted to citrate
- Citrate-pyruvate shuttle carries acetyl CoA out from mitochondria to cytosol (*citrate lyase* in cytosol regenerates acetyl CoA) (see Figure 5.12)
- Fatty acids up to 16C are synthesized in cytosol
- Longer fatty acids are synthesized on cytoplasmic face of endoplasmic reticulum
- Fatty acids are synthesized in the liver, packaged into TAG in lipoproteins, and transported to adipose tissue for storage

 Check your understanding

How is fatty acid metabolism compartmentalized? (*Hint: remember different tissues as well as different organelles, and that you should include synthesis as well as breakdown.*)

5.7 KETONE BODY BREAKDOWN AND SYNTHESIS

Ketone bodies, β-hydroxybutyrate and acetoacetate, act as water-soluble forms of fatty acids. They are the preferred fuel of heart muscle and renal cortex, and can be used by the brain in starvation.

- Produced in mitochondria of liver cells for release into blood stream
- Formed from acetyl CoA, especially from β-oxidation of fatty acids
- Ketone body production regulated by the balance between carbohydrate and fatty acid metabolism in cells

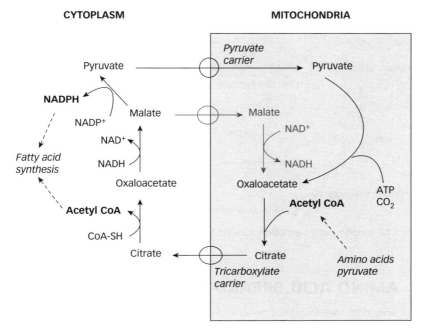

Figure 5.12 Citrate/pyruvate shuttle. Acetyl CoA is combined with oxaloacetate *(citrate synthase)* for export from the mitochondrion. In the cytoplasm, breakdown of citrate regenerates acetyl CoA and, via malate, generates NADPH (for fatty acid synthesis) from NADH. If sufficient NADPH is present in the cytoplasm, malate can be recycled directly via the malate transporter (grey circle, see Figure 5.3).

- **Acetoacetate** synthesis involves the unique enzymes *β-hydroxy β-methylglutaryl CoA (HMG-CoA) synthase*, and *HMG-CoA lyase*, induced in the liver by glucagon on starvation (see Figure 5.13)
- Ketone bodies are catabolized by:
    - i. heart muscle and renal cortex in preference to glucose
    - ii. brain on starvation (FAs are unable to cross blood–brain barrier)
- **Acetone**, a breakdown product of acetoacetate, can be detected on the breath of alcoholics or patients with uncontrolled **diabetes**

Looking for extra marks?

If fatty acid breakdown predominates over carbohydrate breakdown, acetyl CoA in liver forms acetoacetate and β-hydroxybutyrate (via *β-hydroxybutyrate dehydrogenase*). During starvation, fatty acid breakdown generates acetyl CoA, but no oxaloacetate is available (as gluconeogenesis in the liver depletes the mitochondria of oxaloacetate), so acetyl CoA is used instead to generate ketone bodies rather than entering the TCA cycle.

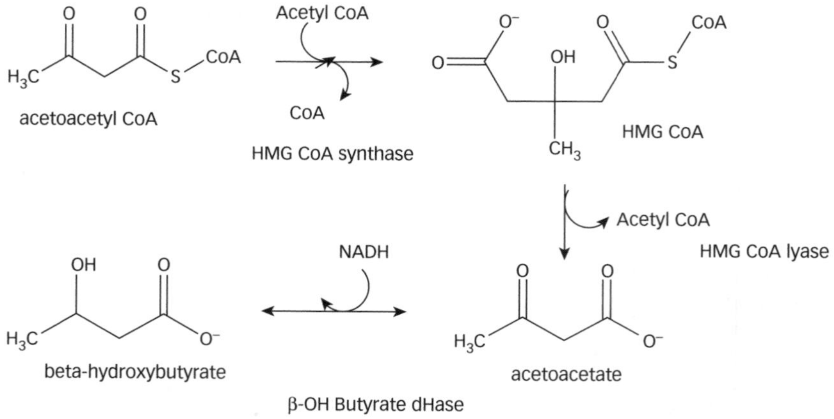

Amino acid breakdown and synthesis

Figure 5.13 Ketone body formation. Ketone bodies are water-soluble derivatives of fatty acids.

5.8 AMINO ACID BREAKDOWN AND SYNTHESIS

Proteins are degraded into peptides or individual amino acids by proteases and peptidases. The catabolism of amino acids involves separation of the amino group from the carbon skeleton. The amino group is finally converted into urea for excretion in mammals, and the carbon skeletons enter other catabolic pathways including gluconeogenesis and ketogenesis.

Protein degradation

- *Proteases* (also known as *proteinases*) are digestive enzymes, especially in intestine, which act to break down dietary proteins into short peptides by hydrolysing peptide bonds
- Cellular proteins to be broken down are marked for degradation by conjugation to **ubiquitin** and progressively broken down into constituent amino acids in the **26S proteasome**
    - i. 20S catalytic component
    - ii. 19S regulatory component
- Major classes of proteases include:
    - i. serine proteases (e.g. *chymotrypsin* and *trypsin* in the mammalian gut, and *elastase* in mammalian skin, which can contribute to wrinkles of ageing)
    - ii. acid proteases (e.g. *pepsin* in the stomach, acting at pH <5)
    - iii. thiol proteases (e.g. *papain* from pineapple)
    - iv. metalloproteases (e.g. *collagenase*)
- Peptides are then degraded to constituent amino acids by *peptidases* that break peptide bonds

Amino acid breakdown

1. How are amino acids broken down?

- Amino groups of many amino acids are transferred to α-ketoglutarate to generate glutamate by *transaminases* (*aminotransferases*)
    - i. e.g. *aspartate aminotransferase*: asp + α-ketoglutarate $\rightarrow$ oxaloacetate + glutamate
    - ii. e.g. *alanine aminotransferase*: ala + α-ketoglutarate $\rightarrow$ pyruvate + glutamate
- Ammonium ion NH_4^+ is formed by oxidative deamination of glutamate by *glutamate dehydrogenase* in a reaction that generates $NADH + H^+$ (or $NADPH + H^+$) (this enzyme is unusual as it can utilize either NAD^+ or $NADP^+$)
- Serine and threonine can be directly deaminated:
    - i. serine $\rightarrow$ pyruvate + NH_4^+
    - ii. threonine $\rightarrow$ α-ketobutyrate + NH_4^+
- In mammals, NH_4^+ is converted into **urea** for secretion via the **urea cycle** (see Figure 5.16)
- In birds and terrestrial reptiles, NH_4^+ is converted into uric acid, while many aquatic animals secrete NH_4^+ directly
- Carbon skeletons are either **glucogenic** or **ketogenic**, or have both potentials (Figure 5.14)

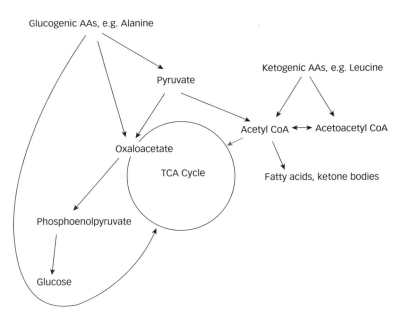

Figure 5.14 Amino acids can be either ketogenic or glucogenic, or both. The carbon skeleton of glucogenic amino acids are broken down to a component of the TCA cycle or pyruvate and can be used to make glucose. The carbon skeleton of ketogenic amino acids are broken down to acetyl CoA. In animals, acetyl CoA cannot be converted back into glucose.

Amino acid breakdown and synthesis

- Ketogenic amino acids (e.g. leucine) give rise to acetyl CoA or acetoacetyl CoA which can be converted into:
  - i. ketone bodies
  - ii. fatty acids, or
  - iii. oxidized in the TCA cycle
  - iv. In animals the inability to convert acetyl CoA into pyruvate means that these skeletons cannot be used to generate glucose
- Glucogenic amino acids (e.g. glutamine) are degraded to pyruvate or intermediates of the TCA cycle and so can be used to generate glucose

2. **Where are amino acids broken down?**

- Many tissues can degrade amino acids (e.g. muscle) but only liver can undergo urea cycle
- Nitrogen from amino acid breakdown must be released into blood and taken up by liver for conversion to urea—via the alanine cycle (see Figure 5.15)
- Formation of NH_4^+ by *glutamate dehydrogenase* and all steps up to the synthesis of citrulline occur in the mitochondrial matrix of liver cells

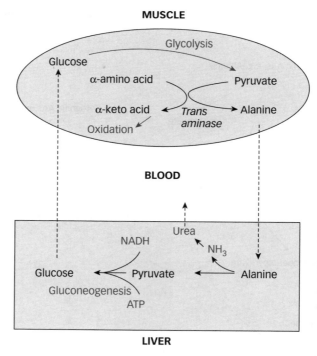

Figure 5.15 Alanine cycle. Amino acids in muscle have their amino groups removed by transfer to pyruvate, allowing oxidation of the carbon chain to provide ATP. The nitrogen is transported (as alanine) in the blood to the liver, where it is removed in the urea cycle and the 3C skeleton converted to glucose, for return to the muscle. In starvation, branched chain amino acids (leu, val) are the major ones used as fuel in muscle.

- Remainder of the **urea cycle** takes place in the cytosol of liver cells
- Urea is released to the blood to be removed by the kidney

3. When are amino acids broken down?

- When an excess of amino acids are ingested in the diet since amino acids cannot be stored. Common in Western diet
- In muscle during prolonged fasting and exercise
- Under extreme starvation when the organism starts to degrade its component parts, especially muscle protein

4. Key examples of amino acid breakdown

- Glucogenic amino acids:
    - i. e.g. alanine can be directly transaminated to pyruvate (direct reversal of synthesis); pyruvate then enters gluconeogenesis (Figure 5.15)
- Ketogenic amino acids:
    - i. e.g. leucine is broken down ultimately to the 2C unit acetyl CoA, which can be converted into ketone bodies, hence it is ketogenic
    - ii. conversion steps include transamination and oxidative decarboxylation to form branched chain acyl CoA molecules (shorter by one C than the α-keto acid precursor)
- Amino acids feeding into both pathways:
    - i. e.g. tyrosine undergoes five reactions (the first being transamination), ultimately forming acetoacetate (a ketone body) plus cytosolic fumarate that is then converted to glucose (i.e. glucogenic)

Looking for extra marks?

Genetic defects in amino acid breakdown pathways can cause major diseases such as phenylketonuria (an accumulation of phenylalanine), which leads to severe mental retardation and death aged 20–30. Treatment is by neonatal screening and a low phenylalanine diet.

Urea cycle

Ammonia produced from amino acid breakdown is highly toxic and so is converted in the urea cycle to a relatively non-toxic compound, urea, for excretion (see Figure 5.16).

- Urea is generated from ammonia in five steps, two in mitochondria, and three in cytosol (compartmentalization allows additional control)
- Mitochondrial steps:
    - i. $NH_4^+ + HCO_3^- + 2ATP \rightarrow$ carbamoyl phosphate $+ 2ADP + P_i$
    - ii. carbamoyl phosphate $+$ L-ornithine $\rightarrow$ citrulline $+ P_i$

Amino acid breakdown and synthesis

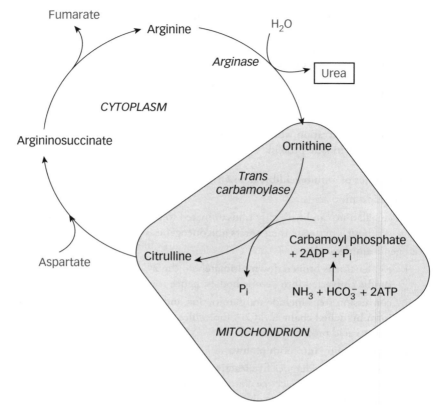

Figure 5.16 Urea cycle. The urea cycle is used to convert ammonia to urea for safe excretion.

- Cytoplasmic steps:
  - i. citrulline + aspartate + ATP → argininosuccinate + AMP + PP_i
  - ii. argininosuccinate → L-arginine + fumarate
  - iii. L-arginine + H_2O → L-ornithine + urea
- L-ornithine then re-enters the mitochondria to repeat the cycle
- First committed step is mediated by *carbamoyl phosphate synthetase I*, an enzyme allosterically activated by *N*-acetyl glutamate, which signals increased glutamate from amino acid breakdown
- All other steps are regulated by concentration of substrate

Amino acid synthesis

Amino acids must be synthesized to allow protein production. Humans are unable to synthesize nine out of 20 amino acids—these nine are essential (Table 5.10) and must be present in the diet. Microorganisms, e.g. *E. coli*, can synthesize all 20.

Essential amino acids	Non-essential amino acids
histidine	alanine
isoleucine	arginine
leucine	asparagine
lysine	aspartate
methionine	cysteine
phenylalanine	glutamate
threonine	glutamine
tryptophan	glycine
valine	proline
	serine
	tyrosine

Table 5.10 Essential and non-essential amino acids in mammals

1. **How are amino acids synthesized?**
- NH_4^+ is assimilated into amino acids via glutamate and glutamine:
  - i. α-ketoglutarate + NH_4^+ + NADPH $\rightarrow$ glutamate + $NADP^+$ + H_2O by *glutamate dehydrogenase*
  - ii. α-ketoglutarate + glutamine + NADPH + H^+ $\rightarrow$ 2 glutamate + $NADP^+$ by *glutamate synthase*
- Reductive power and/or ATP are needed
- Carbon skeletons are made from intermediates of glycolysis, TCA cycle, or pentose phosphate pathway (Table 5.11)
- These intermediates are converted into amino acids by **transamination** reactions
- Some amino acids are synthesized from others (Table 5.11)

2. **Where in the cell are amino acids synthesized?**
- Cytosol

Precursor	Amino acid formed	Can be further converted to
Oxaloacetate	aspartate	asparagine methionine lysine threonine → isoleucine
Pyruvate	alanine valine leucine	
PEP + erythrose 4-phosphate	phenylalanine tryptophan	tyrosine
3-Phosphoglycerate	serine	cysteine, glycine
α-Ketoglutarate	glutamate	glutamine, proline, arginine
Ribose 5-phosphate	histidine	

Table 5.11 Precursors of amino acids (in bacteria and plants)
Essential amino acids in mammals are underlined: these must be taken up in the diet and not synthesized from precursors, unlike the situation in plants and microbes.

3. When are amino acids synthesized?

- When there is a lack in the diet, as they are required for protein synthesis but cannot be stored

4. Regulation of amino acid metabolism

- Biosynthesis (i.e. gene expression) is regulated by feedback inhibition as the major control for these pathways is the availability of amino acids for protein synthesis

Looking for extra marks?

Nitrogen used in amino acid synthesis must enter as ammonia (NH_4^+) and is fixed by microorganisms.

➔ see 8.3 Nitrogen assimilation (p. 215) for nitrogen assimilation in micro-organisms

 ### Check your understanding

How can the carbon skeletons of amino acids be utilized? (*Hint: give an example of a ketogenic and a glucogenic amino acid.*)

What happens to the amine group when amino acids are degraded? (*Hint: remember the alanine cycle as well as the urea cycle.*)

5.9 PENTOSE PHOSPHATE PATHWAY

Key aspects of the pentose phosphate pathway (PPP)

- Also called the phosphogluconate pathway or the hexose monophosphate shunt, the **pentose phosphate pathway** (PPP) serves two primary functions:
    i. production of NADPH reducing power
    ii. synthesis of ribose 5-phosphate (for nucleotide biosynthesis)
- Uses glucose 6-phosphate
- Occurs predominantly in:
    i. tissues that synthesize fatty acids or steroids (e.g. liver, adipose, and adrenal glands)
    ii. rapidly dividing cells (e.g. cancer cells—make lots of DNA)
- Two main stages: oxidative and non-oxidative (see Figure 5.17)
- Takes place in cytoplasm

1. Oxidative stage

- Two-step reaction, overall: glucose 6-phosphate + 2 $NADP^+$ + H_2O → ribulose 5-phosphate + 2 NADPH + CO_2 + $2H^+$
- First reaction is catalysed by *glucose 6-phosphate dehydrogenase*

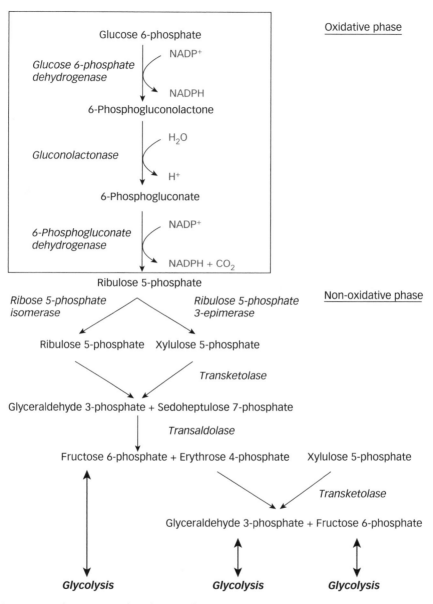

Figure 5.17 The pentose phosphate pathway.

- Second reaction involves ring opening (*lactonase*) followed by oxidative decarboxylation by *6-phosphogluconate dehydrogenase*

2. **Non-oxidative stage**
- Ribulose 5-phosphate usually converted back into glycolytic intermediates

  i. 3 × ribulose 5-phosphate → 2 × fructose 6-phosphate + glyceraldehyde 3-phosphate

  ii. both products (F6P and GA3P) can enter gluconeogenesis or glycolysis

- Alternatively, ribulose 5-phosphate can be converted into other 5C sugar phosphates (xylulose 5-phosphate, ribose 5-phosphate) by *isomerases/epimerases* if high demand for both NADPH and nucleotides

- These are then converted to sugars with different numbers of carbon atoms using:

  i. *transaldolases* which transfer three-carbon units

  ii. *transketolases* which transfer two-carbon units

Looking for extra marks?

Deficiency of *glucose 6-phosphate dehydrogenase*, the first enzyme in the pathway, is an X-linked condition leading to haemolytic anaemia in the presence of some antimalarial or antibiotic drugs.

 Check your understanding

Discuss the central role of acetyl CoA in metabolism. (*Hint: draw a diagram linking FA and carbohydrate metabolism with acetyl CoA, and use this as your essay plan.*)

How is ATP generated aerobically? (*Hint: the ETC is central to answering this question, but also remember to include all sources of acetyl CoA that feed into the TCA cycle and provide electrons for oxidative phosphorylation including beta oxidation of fats.*)

Compare and contrast respiration and photosynthesis. (*Hint: draw the ETCs in both to show the similarities. Note where glucose is on each pathway and the importance of gases O_2 and CO_2, plus co-factors. Remember respiration is catabolic but photosynthesis is anabolic.*)

 see 8.1 Photosynthesis (p. 202) for photosynthetic electron transfer and carbon fixation

5.10 LACTATE AND ETHANOL METABOLISM

Lactate is produced by glycolysis under anaerobic conditions. Ethanol is a highly reduced, two-carbon compound, ingested in the diet, or produced from glucose by some gut bacteria. Both products are toxic and must be broken down. Metabolism of ethanol in the liver involves its oxidation initially to ethanal, and then to acetate.

- Ethanol is produced by some bacteria or fungi by the anaerobic breakdown (**fermentation**) of glucose: pyruvate undergoes decarboxylation and reduction, producing ethanol as a final product

- Ethanol in mammalian bloodstream arises from:

  i. ingestion of alcoholic drinks

  ii. synthesis by gut bacteria

- Levels of ethanol in the bloodstream vary from 0.5 mM (production by gut flora) to 20–100 mM (after heavy drinking).
- Lactate arises from reduction of pyruvate to lactate directly using NADH from glycolysis (cannot be oxidized by O_2)
- Both lactate and ethanol have some toxic characteristics:
  - i. lactate is acidic (can affect enzyme activity) and cannot be effectively oxidized, so gluconeogenesis is inhibited
  - ii. ethanol is lipid-soluble (can affect membrane function)
- In mammals, ethanol is broken down by oxidation to **acetate** by *alcohol dehydrogenase* and *aldehyde dehydrogenase* in the liver

Ethanol breakdown in the liver

- Ethanol + NAD^+ → ethanal + NADH by *alcohol dehydrogenase* in cytoplasm of liver cells
- Ethanal + NAD^+ → acetate + NADH by *aldehyde dehydrogenase* in liver mitochondria
- Acetate converted to acetyl CoA by *acetyl CoA synthetase*:

$$CH_3COO^- + ATP + CoA\text{-}SH \rightarrow CH_3CO\text{-}SCoA + AMP + PP_i$$

- Acetate may be
  - i. used in the liver, or
  - ii. if levels are high, transported to other tissues via the blood
- High levels of alcohol lead (in the short term) to over-reduction of the liver cytoplasmic NAD^+ pool
- Aldehyde group of ethanal is highly reactive:
  - i. e.g. reacts with amino groups of proteins ($RCHO + R'NH_2 \rightarrow$ $R\text{-}CH{=}N\text{-}R' + H_2O$)
  - ii. high levels of, or prolonged exposure to, ethanal can lead to cell damage
- Pathological consequences of long term alcohol exposure include liver damage (cirrhosis, fat deposition in liver)

Lactate metabolism and the Cori cycle

- Lactate is produced by glycolysis under anaerobic conditions:
  - i. in contracting muscle, since rate of pyruvate production is greater than rate of its oxidation in the TCA cycle
  - ii. in red blood cells, as they lack mitochondria so cannot oxidize pyruvate
- Pyruvate + NADH → lactate + NAD^+ catalysed by *lactate dehydrogenase*
- Important pathway to replenish NAD^+ levels for glycolysis in skeletal muscle and red blood cells:

Lactate and ethanol metabolism

    i. allows these cells to continue using glycolysis for energy production

    ii. shifts the metabolic burden to other tissues especially liver

- Lactate diffuses via blood to cardiac muscle or liver:

    i. taken up by cardiac muscle for conversion to pyruvate and generation of ATP through TCA cycle and oxidative phosphorylation (cardiac muscle <u>never</u> functions anaerobically), or

    ii. converted in liver to glucose in gluconeogenesis: lactate → pyruvate → glucose which is then returned to cells requiring it for glycolysis (rbc and contracting muscle) (the **Cori cycle**—see Figure 5.18)

Looking for extra marks?

Lactate dehydrogenase exists as several isoenzymes. It is a tetramer of two different types of protein: H in heart and M in skeletal muscle. H_4 converts lactate to pyruvate in the heart and is inhibited by high pyruvate, while M_4 converts pyruvate to lactate so glycolysis can continue when skeletal muscle is working anaerobically (e.g. sprinting).

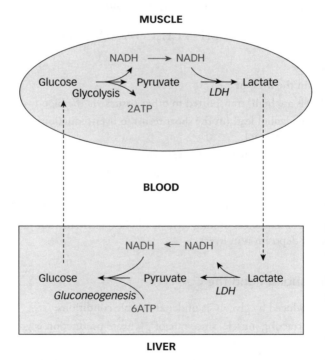

Figure 5.18 Cori cycle. Glucose converted to lactate in anaerobic muscle is carried in the blood to the liver, where it is converted back to glucose and released back into the bloodstream. The muscle gains two molecules ATP/glucose, while the liver uses six ATP/ glucose in gluconeogenesis.

6 Integration of mammalian metabolism

All metabolic pathways are co-ordinated to meet the needs of the whole body as they vary with changes in exercise and nutrition.

Key concepts

- In humans, normal blood glucose levels stay within range 4–6 mM, i.e. relatively constant despite periods of eating and fasting throughout the day
- Minimum viable blood glucose level = 3 mM—tolerated in prolonged starvation
- Glucose is primary fuel for:
    - i. brain (accounts for 60% total glucose utilization in resting state)
    - ii. red blood cells (no mitochondria—glycolysis only)
    - iii. contracting muscle
- Liver is responsible for buffering blood glucose levels—is an altruistic organ, supplying the glucose needs of other organs
- Other usable fuels include:
    - i. fatty acids, e.g. resting muscle
    - ii. ketone bodies and lactate, e.g. heart muscle
    - iii. keto acids from amino acids, e.g. liver

continued

Key organs in metabolic integration

- Carbohydrate stores (glycogen) are exhausted within one day of fasting at basal rate metabolism
- Lipids (triacylglycerols, or TAG) are the major fuel reserve of body, but cannot be transported in blood directly: instead are mobilized as free fatty acids (FFA) which are then bound to albumin, or as ketone bodies (water soluble)
- TAG content varies with individual, but normally lasts about three months fasting
- **ATP, the energy currency of the cell, is not stored (turns over in seconds)**
- Table 6.1 summarizes the body's fuel stores

Storage compound	kJ stored
Glycogen	6,500
Triacylglycerols	550,000
Mobilizable protein	100,000

Table 6.1 Fuel stores in average 70 kg man

6.1 KEY ORGANS IN METABOLIC INTEGRATION

Different tissues (brain, heart, muscle) utilize different chemical fuels, and their requirement will vary with different metabolic demands (e.g. starvation, exercise). The liver and adipose tissue are key organs in metabolic integration, storing or releasing various fuels as required (Figure 6.1).

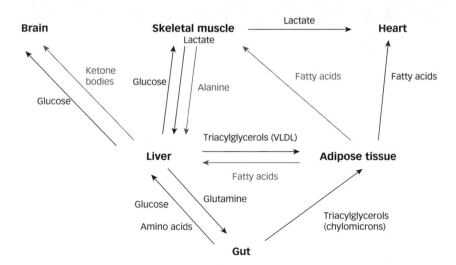

Figure 6.1 Key organs in metabolic integration. Starvation conditions are in grey.

Liver

- Most dietary compounds pass through liver, hence rapid regulation possible
- Altruistic organ: provides glucose and ketone bodies as fuel to brain, muscle, and other organs and tissues
- Uses keto acids from amino acid breakdown as own fuel in preference to glucose
- Glycolysis in liver generates building blocks for biosynthesis NOT for liver's own energy needs
- Liver cannot use acetoacetate as fuel because has low concentration of β-*acyl CoA transferase* needed to catalyse succinyl-CoA + acetoacetate $\rightleftharpoons$ succinate + aceto-acetyl-CoA

1. Glucose

- Liver takes up large amounts of glucose from blood
- Glucose transport into liver is rapid (GLUT-2 transporter) and not insulin-dependent
- Liver-specific form of *hexokinase* (*glucokinase*) has a high K_m (~10 mM) and is not inhibited by glucose-6-phosphate (G6P)—hence liver can respond to raised blood glucose by taking in glucose
- Liver synthesizes and stores glycogen from the glucose taken up from blood
- Releases glucose to blood, derived from:
  - i. gluconeogenesis: alanine and lactate from muscle
    glycerol from adipose tissue
    glucogenic amino acids from diet
  - ii. Glycogenolysis from endogenous glycogen stores
- Possesses *glucose 6-phosphatase*—uses glucose 6-phosphate to generate free glucose which can cross the plasma membrane of hepatocytes and enter blood for transport to other organs
- Liver takes on metabolic burden of contracting muscle by converting lactate from muscle to glucose, which is transported back to muscle—the **Cori cycle**
  - ➜ *see Figure 5.18 (p. 166)*

2. Lipids

- Fed state:
  - i. fatty acids are synthesized from glucose and keto acids, and enter fuel stores
  - ii. *carnitine acyl transferase I* (*CPT1*) is inhibited, so long chain fatty acids stay in cytoplasm
  - iii. fatty acids are esterified into TAG and released to blood in VLDL (**very low density lipoprotein**)
- Fasting state
  - i. fatty acids are used as fuel
  - ii. *CPT1* is active so fatty acids enter mitochondria and are degraded by β oxidation pathway

Key organs in metabolic integration

→ *see 5.6 Lipid breakdown and synthesis (p. 149) for β-oxidation*

  iii. acetyl CoA (formed on β-oxidation of fats) may be further oxidized or converted to ketone bodies

→ *see 5.7 Ketone body synthesis and breakdown (p. 154) for ketone body metabolism*

3. Amino acids

- Amino acids are deaminated to produce keto acids and ammonia
- Ammonia is converted to urea (via **urea cycle**)

  → *see Figure 5.16 (p. 160)*

- Depending on their structure, amino acids can be converted into glucose (**glucogenic** amino acids) or fatty acids (**ketogenic** amino acids)

  → *see 5.8 Amino acid breakdown and synthesis (p. 158)*

Brain

- Lacks fuel stores - relies on liver to provide metabolizable fuels
- Glucose = main brain fuel in normal conditions—requires continuous glucose supply from liver via blood
- Glucose taken up from blood by GLUT1 transporter (largely insulin independent)
- Brain accounts for ~60% of total body's glucose consumption (~120 g/day for average 70 kg male), mainly for ion pumping
- Fatty acids cannot cross blood–brain barrier, but water-soluble ketone bodies can cross
- On starvation, ketone bodies are used by the brain instead of glucose

Muscle

- Fuel store = glycogen (75% of body glycogen is stored in muscle)
- Can use glucose, fatty acids and ketone bodies as fuels
- Glucose is taken up from blood by GLUT4 transporter (insulin-dependent) in fed state
- Glucose released from endogenous glycogen stores (as G6P) stays in muscle cells, since no *glucose 6-phosphatase* is present (unlike liver)
- In rapidly contracting ('fast twitch', 'white') muscle, glycolysis rates are far greater than oxygen supply, so excess pyruvate is converted to lactate or alanine rather than entering TCA cycle
- Lactate is transported to liver for conversion to glucose and transport back to muscle via the Cori cycle

  → *see Figure 5.18 (p. 166) for Cori cycle*

Looking for extra marks?

Different types of muscle prefer different fuels:
   i. resting skeletal muscle uses fatty acids
   ii. contracting fast twitch skeletal muscle uses endogenous glycogen
   iii. contracting red skeletal muscle uses blood glucose or fatty acids
   iv. cardiac muscle uses lactate and ketone bodies

Adipose tissue

- Fuel store = triacylglycerols (TAG), composed of glycerol and three fatty acid chains:
  - *see 1.3 Lipids (p. 12) for TAG structure*
    i. fatty acids synthesized in liver are transported to adipose tissue as TAG in VLDL
    ii. TAG from the diet are transported from the intestine to adipose tissue in **chylomicrons**
- *Lipoprotein lipase* on surface of adipose tissue liberates fatty acids from VLDL/chylomicrons for transport across membrane of adipose cell (adipocyte)
- Once inside adipocyte, fatty acids are activated and CoA derivatives are transferred to glycerol-3-phosphate to reform TAG
    i. glycerol 3-phosphate in adipose is derived from glucose via glycolysis (cf. liver can make glycerol 3-phosphate from glycerol via *glycerol kinase*)
    ii. hence glucose uptake is required for synthesis of TAG in adipose tissue
- **Glucagon** signals when fuel stored in fat is required by other tissues: results in activation of *hormone-sensitive lipase* which hydrolyses TAG into fatty acids:
    i. fatty acids cross out of adipocyte and are transported in blood, bound to albumin
    ii. glycerol released on lipolysis is transported to liver and can be converted back to glucose

Cells of the immune system

- Lymphocytes and enterocytes use circulating glutamine as energy source
- Glutamine is released to the blood from liver and muscles
- Key enzyme in this pathway is *glutaminase*

 Check your understanding

How can the liver buffer blood glucose levels? (*Hint: remember glucokinase, liver-specific version of hexokinase, GLUT2 transporters, large glycogen stores and presence of G6Pase.*)

continued

> What is the major function of adipose tissue? (*Hint: fat storage as TAGs, with release when body needs energy; don't forget that fat also serves as insulation, e.g. in polar bears, and brown adipose generates heat through uncoupling proteins.*)
>
> ➔ *see 5.6 Lipid synthesis and breakdown (p. 147) for fatty acid metabolism*

6.2 KEY METABOLITES IN METABOLIC INTEGRATION

Glucose 6-phosphate, pyruvate, and acetyl CoA are crucial at junctions between many different metabolic pathways, providing a regulatory link between disparate pathways.

Glucose 6-phosphate

- Figure 6.2 summarizes the routes into and out of glucose 6-phosphate in metabolism
- Glucose is immediately phosphorylated to glucose 6-phosphate (G6P) by *hexokinase* on entry to cells (there is virtually no free intracellular glucose)
- *Hexokinase* :
    i. muscle isoenzyme has a low K_m (high affinity) so as to be insulated from changes in blood glucose levels (can still use glucose even when low concentrations in the blood)
    ii. liver isoenzyme (*glucokinase*) has a high K_m (low affinity) for glucose so that it can respond to blood glucose levels (is not saturated by high concentrations of glucose, e.g. after a meal, and is not inhibited by product G6P)

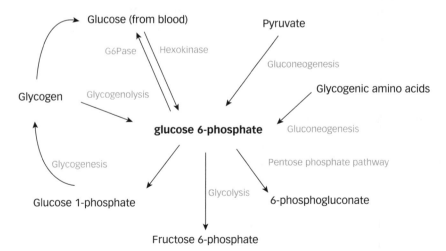

Figure 6.2 Glucose 6-phosphate is central to many pathways of mammalian metabolism. The pathways in which it is involved are in grey and the products or precursors in black.

- In liver and muscle, glucose 6-phosphate can be converted into glycogen which acts as a store of glucose

  ➔ *see 5.5 Glycogen breakdown and synthesis (p. 145) for glycogen synthesis*
- Liver and kidney are unique in having the enzyme *glucose 6-phosphatase*, so can release glucose into blood from intracellular glucose 6-phosphate when blood glucose levels drop
- Enzymes for gluconeogenic pathways are mainly found in liver, and induced by starvation

  ➔ *see 5.2 Glucose breakdown and synthesis (p. 134) for gluconeogenesis*
- Most amino acids (alanine, glutamate, serine, histidine, etc.) are glucogenic, i.e. they can provide the carbon skeletons for glucose synthesis. Leucine and lysine cannot—they are ketogenic

  ➔ *see 5.8 Amino acid breakdown and synthesis (p. 158) for glucogenic and ketogenic amino acids*

Pyruvate

- Figure 6.3 summarizes the routes into and out of pyruvate in metabolism
- Pyruvate is formed in glycolysis from:
  - i. glucose
  - ii. amino acids after deamination
  - iii. blood lactate
- Pyruvate is converted to lactate in fast twitch muscle and erythrocytes; lactate is converted back to pyruvate in heart and liver

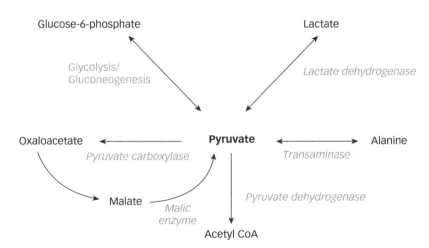

Figure 6.3 Pyruvate is central to many pathways of mammalian metabolism. The pathways in which it is involved are in grey and the products or precursors in black.

Key metabolites in metabolic integration

- Different isoenzymes of *lactate dehydrogenase* are adapted for the two different directions:
    i. heart isozyme H_4 converts lactate $\rightarrow$ pyruvate
    ii. muscle isozyme M_4 converts pyruvate $\rightarrow$ lactate
- Pyruvate enters mitochondria and is converted to acetyl CoA by oxidation + decarboxylation (**oxidative decarboxylation**) catalysed by *pyruvate dehydrogenase*
- This is the '**link reaction**' between glycolysis and the TCA cycle
- *Pyruvate dehydrogenase* reaction is irreversible in mammals and is controlled by phosphorylation, $NAD^+/NADH$ levels and acetyl CoA/CoA ratios
- Pyruvate (made from lactate, alanine, or glycerol) is a precursor for gluconeogenesis in the liver/kidney
- *Pyruvate kinase* reaction (formation of pyruvate in glycolysis)

  is irreversible, and must be bypassed in gluconeogenesis by the *pyruvate carboxylase/ phosphoenolpyruvate carboxykinase* (PEPCK) enzymes

  ➔ *see Figure 5.2 (p.132)*

Acetyl CoA

- Figure 6.4 summarizes the routes into and out of acetyl CoA in metabolism
- Acetyl CoA can be derived from:
    i. pyruvate (from glycolysis)
    ii. breakdown of fatty acids
    iii. ketogenic amino acids
- Acetyl CoA is oxidized via the TCA cycle as a major source of ATP in cells

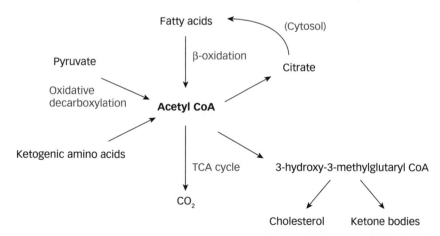

Figure 6.4 Acetyl CoA is central to many pathways of mammalian metabolism. The pathways in which it is involved are in grey and the products or precursors in black.

- Acetyl CoA is a precursor for fatty acid synthesis, but (in mammals) cannot be converted back into glucose
- Fatty acid synthesis takes place in the cytoplasm, fatty acid oxidation in the mitochondria
- Acetyl CoA must be activated for fatty acid synthesis by carboxylation to malonyl CoA. This carboxylation requires a biotin cofactor and ATP + CO_2
- Acetyl CoA is a precursor for HMG-CoA. This is an intermediate in cholesterol biosynthesis (inhibited by statins) and other steroid derivatives
- In liver, HMG CoA can be converted into ketone bodies which provide fuel for the brain and heart on starvation

Looking for extra marks?

Acetyl CoA is used in modification of proteins (e.g. histone acetylation) and synthesis of smaller molecules (e.g. acetylcholine).

 Check your understanding

How can acetyl CoA be used to provide fuels for other tissues? (*Hint: remember acetyl CoA cannot be converted back to pyruvate, and therefore glucose, in animals, but it is a major precursor for fatty acid synthesis.*)

What are the potential fates of pyruvate? (*Hint: remember it can be used in anabolic as well as catabolic pathways.*)

6.3 KEY ASPECTS OF FUEL UTILIZATION

Average day with periods of eating and fasting

- On feeding, blood glucose levels rise, and glucose is taken up into the liver (via GLUT2 transporters and the high K_m *hexokinase*) for storage as glycogen or conversion to fats
- Pancreas secretes **insulin** which leads to uptake of glucose (via GLUT4 transporters) by other tissues, especially:
  - i. muscle (where it is stored as glycogen)
  - ii. adipose tissue (where it forms glycerol 3-phosphate for fatty acid esterification)
- On fasting, blood glucose levels tend to fall, and **glucagon** levels rise in response, leading to:
  - i. mobilization of glucose from glycogen stores in the liver
  - ii. lipolysis of TAG and release of fatty acids from adipose tissue
- Changes in fuel utilization with nutritional state are summarised in Table 6.2

Key aspects of fuel utilization

	Fed state (high glucose)	Fasting state (low glucose)
Insulin (β cells, pancreas)	⇑	⇓
Glucagon (α cells, pancreas)	⇓	⇑
Liver	⇑ glucose uptake (insulin independent) GLUT2 ⇑ glycogen synthesis	glucose release by glycogenolysis which yields glucose 6-phosphate and glucose
Muscle	⇑ glucose uptake (insulin dependent) GLUT4 ⇑ glycogen synthesis	⇓ glucose uptake since low insulin ⇓ glucose utilization
Adipocytes	⇑ glucose uptake ⇑ formation glycerol 3-phosphate and hence incr. TAG synthesis	⇓ glucose uptake since low insulin ⇑ TAG breakdown → FA

Table 6.2 Comparison of fuel utilization in fed and fasting states

Acute exercise (sprint)

- Anticipation: adrenaline stimulates cAMP cascade in muscle
    i. *PKA* activated
    ii. *glycogen synthase* inhibited
    iii. *glycogen phosphorylase* activated
    iv. overall result: glycogenolysis, generating glucose 6-phosphate
- During: increase in AMP levels signals low energy and stimulates glycogenolysis (as above), and glycolysis via *phosphofructokinase*
    ➔ *see 5.2 Glucose breakdown and synthesis (p. 132)*
- Glycolysis outstrips TCA cycle in muscle so pyruvate is funnelled into lactate (by *lactate dehydrogenase*) or alanine
- Lactate and alanine are transported to liver for conversion to glucose, which is transported back to muscle via Cori cycle. This removes reducing equivalents from muscle
    ➔ *see Figure 5.18 (p. 166) for Cori cycle*
- Net effect: glycogen in muscle is broken down to yield glucose by fast twitch muscle
- Liver needs to continue to oxidize excess lactate after exercise ceases ('oxygen debt')
- Hence, liver bears brunt of metabolism of exercising muscle in sprint
- NB In prolonged exercise (e.g. long distance running), red muscles use oxidative metabolism to completely oxidize glucose and fatty acids
    ➔ *see Figure 6.5 (p. 179)*

24-hour fast

- Carbohydrate reserves (glycogen and glucose) are exhausted within 24 hours of fasting
- Blood [glucose] remains >4 mM

- Lipids cannot be converted to carbohydrate in mammals, therefore it is important to limit glucose utilization by shifting catabolism from glucose to fatty acids (or derivatives)
- Three main ways to control blood glucose levels during fasting:
    i. mobilization of glycogen stores and release of glucose by liver in first 24 hours
    ii. release of fatty acids by adipose tissue
    iii. shift in fuel usage of muscle from glucose to fatty acids
- Low blood glucose levels lead to increased glucagon and decreased insulin levels
    ➔ *see 6.4 Key hormones in metabolic integration (p. 178) for more details on hormone action*
- Adipose tissue mobilizes TAG to give fatty acids
- Increased acetyl CoA and citrate levels inhibit glycolysis, i.e. less glucose is broken down in liver and muscle
- Muscles take up less glucose since lower insulin levels (insulin-dependent GLUT4 transporters), but take up fatty acids freely
- β-Oxidation of fats in muscle also inhibits conversion of pyruvate to acetyl CoA by inhibiting *pyruvate dehydrogenase*
- Liver generates glucose by gluconeogenesis and releases glucose to blood (liver derives its own energy from fatty acids, or from keto acids on amino acid breakdown). Glycolysis in liver is inhibited via effect of glucagon on liver *pyruvate kinase*
- Hence less glucose is removed from the blood (by muscle) and more is added (by the liver)
- More glucose is available for tissues/organs that are unable to use fuels other than glucose (e.g. red blood cells) or that preferentially use glucose (e.g. brain)
- Limited proteolysis of tissue (75 g/day in early starvation) provides three-carbon precursors (alanine) for glucose synthesis in liver
- Alanine is exported from muscle to liver where it is converted to glucose

Prolonged starvation

After three days of starvation
- Ketone bodies (acetoacetate and β-hydroxybutyrate) are synthesized in liver from acetyl CoA
- Ketone bodies are released into blood
- ~30% of brain's energy needs are met by ketone bodies
- Muscle can use **branched chain amino acids** (BCAA) from protein breakdown as energy source

After several weeks
- Ketone bodies = major fuel for brain
- Brain glucose use drops from 120 to 40 g/day

- Proteolysis is lower—approximately ~20 g/day cf. 75 g/day in earlier starvation
- Muscle proteins are not major sources of fuel except under extreme starvation
- i.e. fat mobilization and shift in fuel use by muscle and brain maintains blood glucose levels at ≥3 mM and preserves protein

Diabetes

- Type 1: insulin not produced (juvenile onset, autoimmune destruction of β-cells in pancreas)
- Type 2: insulin insensitivity (mature onset, obesity-related)
- Skeletal muscle, heart and adipose tissue cannot take up glucose (insulin-dependent GLUT4 carriers are inactive)
- Prolonged raised blood glucose after meals can give rise to:
    i.   excretion of glucose in urine
    ii. non-enzymatic glycosylation (glycation) of proteins and damage/rapid turnover of proteins, especially kidney (diabetic kidney damage) and eye lens (diabetic blindness)
- Free fatty acids are released from adipose tissue
- Ketone bodies and VLDL are released from liver (ketoacidosis)
- Muscles and brain switch to lipid-based fuels (see prolonged starvation)
- Low blood glucose → **hypoglycaemic** coma: results from overtreatment with insulin

 Check your understanding

Which energy stores are used during acute exercise? (*Hint: remember muscle glycogen stores and what triggers their breakdown; also consider what happens to the excess pyruvate produced.*)

What is the difference between type 1 and type 2 diabetes? (*Hint: consider age of onset and biochemical defect—either inability to make insulin, or inability to respond to it; think about how the two diseases are treated in the clinic and why different therapies are needed.*)

6.4 KEY HORMONES IN METABOLIC INTEGRATION

Metabolic integration between tissues is achieved by hormones circulating in the blood. Glucagon and insulin are produced in the pancreas in response to blood glucose levels. Other hormones signal potential metabolic demand (adrenaline) or the level of stored fuels (leptin, ghrelin).

Glucagon

- Polypeptide (active product 28 amino acids cleaved from larger precursor) produced by α-cells of islets of Langerhans in pancreas

- Signals starvation—acts to increase fuel levels in blood
- Binds to **G protein**-coupled receptor (GPCR) on membrane of target cell and signal is relayed by *protein kinase A* (Figure 6.5)
- **Acute** effects (fasting):
  - i. stimulates *hormone-sensitive lipase* in adipose tissue (releases free fatty acids)
  - ii. promotes liver glycogenolysis (releases glucose from liver)
  - iii. inhibits glycolysis in liver (by inhibition of *phosphofructokinase 2* and *pyruvate kinase*)
- **Chronic** effects (starvation):
  - i. switches on transcription of *glucose 6-phosphatase* in liver
  - ii. switches on transcription of *hydroxymethylglutaryl CoA lyase* (promotes ketone body biosynthesis)

Adrenaline

- Amino acid derivative produced in adrenal cortex
- Signals 'fight, flight', and 'fancy' (i.e. sexual arousal)—acts to increase fuel levels in blood
- Binds to G protein-coupled receptor and signal is relayed by *protein kinase A* (see Figure 6.5)—pathway very similar to that used by glucagon

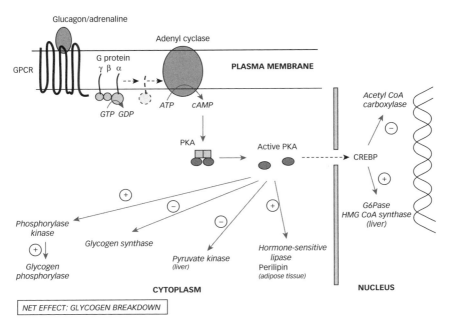

Figure 6.5 Glucagon/adrenaline signalling through GPCR and *protein kinase A* (PKA) leads to release of fuels into the circulation and reduction of genes.

Key hormones in metabolic integration

- Acute effects:
    i. promotes muscle glycogenolysis (releases glucose phosphate into muscle cytoplasm)
    ii. stimulates *hormone-sensitive lipase* in adipose tissue (releases free fatty acids)
    iii. promotes liver glycogenolysis (liver releases glucose into blood)
- Overrides insulin when both are high:
    i. e.g. when you need to escape a predator after a heavy meal
    ii. but: adrenaline stimulates insulin release, as insulin-dependent GLUT4 glucose transporters are needed for muscle uptake of the increased levels of glucose in the blood resulting from adrenaline action

Insulin

- Polypeptide (51 amino acids) produced by β-cells of islets of Langerhans in pancreas from proinsulin precursor that is cleaved
- Signals fed state—acts to increase fuel storage in tissues
- Released from secretory vesicles in β-cells in response to high blood glucose:
    i. high blood glucose $\rightarrow$ uptake of glucose into β-cells via GLUT2 transporters
    ii. glucose used in glycolysis, TCA cycle, and oxidative phosphorylation $\rightarrow$ ATP generated
    iii. high ATP triggers closure of ATP-dependent K^+ channels leading to plasma membrane depolarization
    iv. depolarization opens voltage-gated Ca^{2+} channels
    v. increase in intracellular Ca^{2+} stimulates *phospholipase C (PLC)*
    vi. *PLC* generates IP_3 + DAG from membrane phosphatidyl inositol
    vii. IP_3 stimulates Ca^{2+} release from endoplasmic reticulum
    viii. high intracellular Ca^{2+} triggers release of insulin from secretory vesicles to blood
- Insulin in blood binds to *receptor tyrosine kinase (RTK)* on membrane of target cell and signal is relayed by *protein kinase B (PKB, also known as Akt)* (see Figure 6.6)
- Major acute effects:
    i. stimulates GLUT4 transporter, and hence uptake of glucose into muscle
    ii. stimulates re-esterification of free fatty acids in adipose tissue to form TAG for storage
    iii. promotes glycogen synthesis in muscle and liver (stores glucose)
    iv. inhibits gluconeogenesis in liver
- Other metabolic effects of insulin:
    i. inhibits proteolysis and lipolysis
    ii. increases amino acid uptake from the blood
    iii. decreases autophagy
    iv. increases HCl secretion in the stomach
- Chronic effect: promotes protein synthesis and cell growth

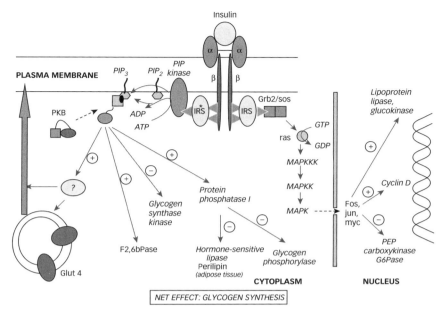

Figure 6.6 Insulin signalling through RTK and protein kinase B, in the fed state.

Hormonal regulation of glycogen metabolism

Reciprocal regulation of catabolic and anabolic pathways is typified by regulation of glycogen synthesis and degradation by insulin/glucagon (Figure 6.7).

- Glucagon (starvation) stimulates *protein kinase A* (*PKA*)
  - i. activates *glycogen phosphorylase kinase* → activates *glycogen phosphorylase*
  - ii. activates *glycogen synthase kinase* → inhibits *glycogen synthase*
  - iii. net effect: promotes glycogen breakdown
- Adrenaline acts similarly to glucagon to stimulate glycogen breakdown and increase blood glucose levels
- Insulin (fed state) stimulates *protein kinase B* (*PKB*) (also known as *Akt*):
  - i. inhibits *glycogen synthase kinase* → activates *glycogen synthase*
  - ii. indirectly activates *protein phosphatase 1* (*PP1*), which dephosphorylates and inactivates *glycogen phosphorylase*
  - iii. also activates ***phosphodiesterase***, which breaks down cAMP to AMP hence removing signal that activates *PKA*
  - iv. net effect: promotes glycogen synthesis, and lowers blood glucose levels

Leptin

- Polypeptide (167 amino acids) produced in adipose tissue
- Signals satiety after feeding and amount of stored triglyceride (adiposity)

Key hormones in metabolic integration

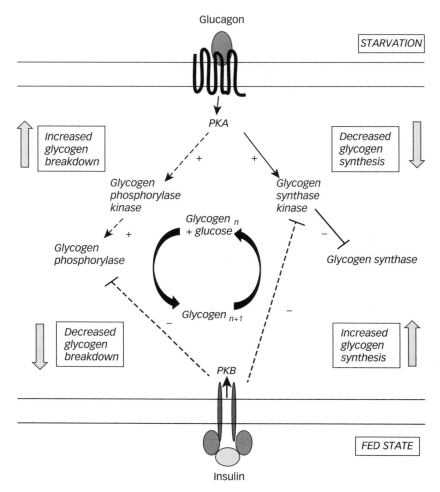

Figure 6.7 Hormonal regulation of glycogen metabolism. Glucagon stimulates formation of cAMP by *adenyl cyclase*, leading to activation of *PKA*, which phosphorylates and activates *glycogen phosphorylase kinase*, which in turn activates *glycogen phosphorylase* by phosphorylating it. *PKA* activates *glycogen synthase kinase* which phosphorylates and inactivates *glycogen synthase*. Hence the net effect of glucagon signalling is to increase breakdown of glycogen to mobilize glucose. By contrast, insulin indirectly activates *protein phosphatase 1 (PP1)* which dephosphorylates and inactivates *glycogen synthase kinase* and *glycogen phosphorylase*. In addition, insulin leads to activation of *phosphodiesterase* that converts cAMP to AMP, hence downregulating the second messenger that activates *PKA*. The net effect of insulin signalling is thus to promote glycogen synthesis.

- Binds to *receptor tyrosine kinase* and signal is relayed by *AMP kinase*
- Acts in hypothalamus to suppress appetite by decreasing release of neuropeptide Y
- Deficiency of leptin gives severe obesity in mice (ob$^{-/-}$ mice)

Ghrelin

- Oligopeptide (28 amino acids) produced in ε-cells of islets of Langerhans in pancreas
- Active form has octanoic acid bound to one of its amino acids
- Signals hunger
- Binds to a G-protein-coupled receptor and signal is relayed via *AMP kinase*
- Acts in hypothalamus to promote appetite by increasing release of neuropeptide Y (i.e. opposite to leptin)
- High levels of ghrelin found in Prader–Willi syndrome, accompanied by excessive appetite

6.5 PRINCIPLES OF HORMONE SIGNALLING IN METABOLISM

Water-soluble hormones in the blood bind to receptors on target cells. Conformational changes in the receptor trigger a response in the target cell. Effects may be rapid ('acute'), where enzymes are switched on/off directly, or longer term ('chronic') via activation of transcription of genes.

Key concepts in hormonal signalling

- Two main classes of receptor:
    - i. G-protein-coupled receptors (e.g. for adrenaline and glucagon)
    - ii. Receptor *tyrosine kinases* (e.g. for insulin and growth factors)
- Receptor activation stimulates production of **second messengers**—diffusible molecules inside the cell which transmit the signal to intracellular components, e.g.
    - i. cAMP, which is synthesized from ATP
    - ii. Ca^{2+}, which is released from intracellular stores (especially endoplasmic reticulum) or enters from outside cell
- Often a cascade of *protein kinases* amplifies the effect of a few activated receptor molecules by phosphorylating downstream target proteins

---

G protein-coupled receptors (GPCR)

- G-protein-coupled receptors have seven transmembrane helices and relatively small extramembrane regions but no enzyme activity
- Human genome encodes >800 GPCR (3.5% of genome)
- Hormones binding to GPCR include:
    - i. glucagon
    - ii. adrenaline

    iii. vasopressin

    iv. prostaglandins

1. GPCR responses involving *protein kinase A*

- *cAMP-dependent protein kinase* or *protein kinase A* (*PKA*) is activated by cyclic AMP (**cAMP**)
- The sequence of events induced by glucagon is typical (Figure 6.5)
- Hormone (glucagon) binding extracellular domain changes conformation of GPCR
- Ligand-bound GPCR stimulates intracellular GDP to GTP exchange on α subunit of **heterotrimeric** G protein
- Largest subunit G_α dissociates from the other two subunits ($G_{\beta\gamma}$) and activates *adenyl cyclase*, stimulating it to produce cAMP (second messenger) from ATP
- cAMP diffuses into cytoplasm and activates *protein kinase A* (*PKA*)
- Activated *PKA* phosphorylates various target enzymes changing their activity:
  - i. ON: *phosphorylase kinase, hormone-sensitive lipase*
  - ii. OFF: *glycogen synthase, pyruvate kinase*
- Activated *PKA* in nucleus phosphorylates <u>c</u>AMP-<u>r</u>esponse <u>e</u>lement <u>b</u>inding protein (CREBP), which then activates transcription of gluconeogenic genes (e.g. *glucose 6-phosphatase*) and inhibits transcription of genes involved in lipogenesis
- Downregulation of cAMP response by:
  - i. *protein phosphatases* (constitutive; also indirectly activated by insulin) dephosphorylate target proteins, e.g. *PP1* dephosphorylates *glycogen phosphorylase*
  - ii. cAMP hydrolysed by *cAMP-phosphodiesterase* (phosphodiesterase is activated indirectly by insulin)
  - iii. GTP is slowly hydrolysed by G_α, converting it back to the inactive GDP-bound form, which then re-associates with $G_{\beta\gamma}$ to reform the heterotrimer
  - iv. Other hormones (e.g. somatostatin) release an inhibitory $G_{i\alpha}$ from their heterotrimeric G protein to inhibit *adenyl cyclase*
  - v. On prolonged stimulation, $G_{\beta\gamma}$ promotes uptake of GPCR into the cell by **endocytosis**
- Since *PKA* is common to a large number of signalling pathways, different responses in different tissues are achieved by:
  - i. tissue-specific GPCRs, e.g. glucagon receptors on liver and adipose cells; somatostatin receptors mainly on gut tissue
  - ii. tissue-specific targets for PKA, e.g. *hormone-sensitive lipase* is restricted to adipose tissue, liver-specific isoenzyme of *pyruvate kinase* inhibited by *PKA* while the muscle isoenzyme is not a substrate

2. GPCR responses involving *protein kinase C*

- The sequence of events induced by vasopressin (acting largely in the kidney) is typical (Figure 6.8)

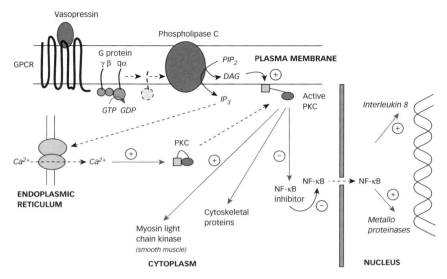

Figure 6.8 Vasopressin signalling through GPCR and *protein kinase C*.

- Hormone (e.g. vasopressin) binding to target GPCR induces GDP/GTP exchange on α-subunit of a heterotrimeric G protein
- $G_{q\alpha}$ activates *phospholipase C* to generate inositol trisphosphate (IP_3) from the membrane phospholipid, phosphatidylinositol bisphosphate (PIP_2)
- Second messenger IP_3 diffuses into cytoplasm and stimulates release of Ca^{2+} from ER
- Ca^{2+}, together with diacylglycerol (other product of PIP_2 hydrolysis) activates *protein kinase C* (*PKC*)
- Targets for *PKC* include cytoskeletal proteins (alter cell shape) and transcription factors (e.g. NF-κB)
- Response is switched off by:
  - i. GTP hydrolysis by $G_{q\alpha}$
  - ii. DAG hydrolysed by *phospholipase A2*
  - iii. IP_3 phosphorylated to inactive IP_4 by *IP_3 kinase*

Revision tip

Protein kinase $\underline{A}$ is activated by c$\underline{A}$MP; *protein kinase $\underline{C}$* is activated by $\underline{C}a^{2+}$ and DAG.

Receptor tyrosine kinases (RTKs)

- Protein domains of receptor include:
  - i. extracellular ligand binding
  - ii. single pass transmembrane helix
  - iii. intracellular *tyrosine kinase* domain

Principles of hormone signalling in metabolism

- *RTKs* are usually dimeric in their active form:
    i. insulin receptor is a permanent dimer
    ii. receptors for other growth factors (e.g. PDGF, EGF) are monomeric when inactive and dimerize only when the hormone binds
- Some receptors (e.g. human growth hormone) lack a *tyrosine kinase* domain and recruit (and activate) a *tyrosine kinase* (*JAK*) only on hormone binding
- Hormone binding triggers *trans*-**autophosphorylation** of tyrosine residues on intracellular region of receptor (hence need for dimerization)
- Phosphotyrosine residues are recognized by 'adaptor' proteins containing SH2 (Src homology 2) domains, which relay the signal
- Many *RTKs* bind growth hormones, and are often implicated in tumour formation
- Insulin receptor activates two independent signalling pathways (see Figure 6.6):
    i. activation of *protein kinase B* (*PKB*)
    ii. small (i.e. monomeric) G protein *ras*—results in activation of the *MAP kinase* pathway; limited period of action due to GTP hydrolysis cf. heterotrimeric G proteins

---

Looking for extra marks?

Mutations of *ras* that prevent GTP hydrolysis result in constitutive *ras* signalling which can be carcinogenic, e.g. *ras* G12V mutation is associated with bladder cancer.

---

1. PKB pathway

- *Protein kinase B* is activated by binding to the phospholipid PIP_3. It is responsible for most of the <u>acute</u> effects of insulin
- Hormone-binding extracellular domain changes conformation of *RTK*
- Ligand-bound *RTK* phosphorylates itself (trans-autophosphorylation) and intermediate proteins, **insulin receptor substrates** (IRS), on tyrosine residues.
- Phosphorylated IRS binds *phosphatidylinositol 3′ kinase* (*PI3K*) bringing it to membrane, where it phosphorylates PIP_2 forming PIP_3
- *PKB* binds PIP_3 at the membrane where it is phosphorylated on serine/threonine residues, leading to activation
- Activated *PKB* phosphorylates various target enzymes, e.g.
    i. ON—*protein phosphatase 1*
    ii. OFF—*glycogen synthase kinase*, ultimately stimulating glycogen synthesis
- Activated *PKB* phosphorylates a protein (still unidentified) which promotes GLUT4 transport to the cell membrane from internal stores

2. *MAP kinase* pathway

- MAP kinase cascade is responsible for <u>chronic</u> effects of insulin (and of many other growth factors)

- Insulin binding to its *RTK* leads to autophosphorylation of receptor and phosphorylation of IRS on tyrosine residues
- Adaptor protein Grb2 recognizes the phosphotyrosine through its SH2 domain, and activates GTP for GDP exchange on *ras*, a monomeric G protein
- *Ras* activates a serine/threonine protein kinase, *MAP kinase 3* (*MAPK3*)
- *MAPK3* then phosphorylates and activates a second kinase, *MAPK2*, which in turn phosphorylates and activates a third kinase (*MAPK*)
- At each stage, a few molecules of one kinase will phosphorylate a large number of the next, hence amplification through a cascade
- Activated *MAPK* has various targets, e.g. transcription factors (myc, jun, fos), switching on genes, e.g. for lipogenesis

 Check your understanding

How does a heterotrimeric G protein relay a signal? (*Hint: remember it is also involved in turning the signal off by hydrolysing GTP to GDP.*)

What is a kinase cascade? (*Hint: think about the MAP kinase cascade triggered by insulin binding to its RTK; remember that such cascades result in signal amplification so a few hormone molecules binding to surface receptors can result in a very marked cellular response, both in acute and chronic phase; make sure you can describe named examples of ligands, receptors, and all the kinases in a cascade.*)

Discuss the role of acetyl CoA in metabolism. (*Hint: start with Figure 6.4 and make sure you include details of all the pathways and components.*)

How is glycogen metabolism regulated by hormones? (*Hint: start with Figure 6.7 and make sure you describe each step. Remember to include also the effect of adrenaline on glycogen metabolism.*)

Looking for extra marks?

The insulin signalling pathway is very similar to that used for growth factor signalling leading to cell proliferation, e.g. epidermal growth factor (EGF), platelet-derived growth factor (PDGF), etc. Growth factors bind to their cognate *RTKs*, which leads to activation of *PLC* and release of IP_3 and DAG, followed by activation of *PKC* and a range of downstream kinases, often including *MAP kinase*. The endpoint is activation of transcription factors that transcribe genes whose products promote passage through the cell division cycle, and of enzymes that restructure the cytoskeleton to permit mitotic spindle formation and cell division.

7 Metabolism of structural components

7.1 NUCLEOTIDES

Key concepts: nucleotides

- Within cells, **purines** (Pu) and **pyrimidines** (Py) exist as nucleotides and polynucleotides
- Nucleotides form:
    i. long polymers of RNA and DNA
    ii. energy currency of the cell (ATP)
    iii. co-substrates (e.g. NAD)
    iv. regulators (e.g. cAMP)
- They can be synthesized *de novo*, or via salvage pathways that recycle pre-existing components such as RNA to make new nucleotides, e.g. for DNA synthesis
- Dietary nucleotides are broken down and products used for energy generation, with excretion of the nitrogenous components
  - ➜ *see 1.5 DNA (p. 19) and 1.6 RNA (p. 26) for more details on structure of nucleotides and polynucleotides*

Breakdown of nucleotides

- Cells break down nucleotides:
    - i. in dietary excess of purine or pyrimidine
    - ii. purines are cycled during muscle contraction to provide precursors for TCA cycle (see Figure 7.1)
- Inosine monophosphate is an intermediate in purine breakdown
- End products of purine catabolism:
    - i. uric acid: birds, some reptiles, and primates
    - ii. allantoin: most mammals (not primates), turtles, some insects, etc.
    - iii. allantoate: some bony fish
    - iv. urea: most fish, amphibians, and fresh water molluscs
    - v. $CO_2 + NH_3$: plants, crustaceans, and many marine invertebrates
- End products of pyrimidine catabolism enter TCA cycle:
    - i. cytidine → uridine → ribose-1-phosphate + uracil →→→ acetyl CoA
    - ii. thymine →→→ succinyl CoA
- Ammonia and bicarbonate are also formed from pyrimidine breakdown

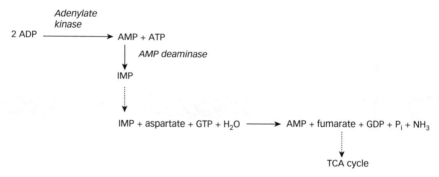

Figure 7.1 Nucleotide breakdown.

Biosynthesis of nucleotides

- Almost all organisms can synthesize ribonucleotides and deoxyribonucleotides:
    - i. *de novo* (for ribonucleotides, e.g. in RNA): activated ribose + base → nucleotide
    - ii. via salvage pathway (uses intermediates from nucleotide breakdown to synthesize new nucleotides): activated ribose + amino acids + ATP + CO_2 → nucleotide
    - iii. not all tissues can synthesize nucleotides *de novo*, so use salvage pathway instead
- C and N atoms of purines and pyrimidines are derived from a few key compounds (see Figure 7.2): glutamine, glycine, aspartate

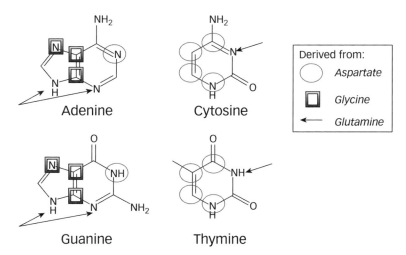

Figure 7.2 Derivation of C and N atoms in pyrimidines (C and T) and purines (G and A) from three key precursors—aspartate, glycine, and glutamine.

- Activated ribose PRPP (5-phospho-α-D-ribosyl 1-pyrophosphate) is key molecule acting as start point for purine and pyrimidine synthesis: PRPP is generated from ribose 5-phosphate by *PRPP synthetase*
  - ➔ *see 5.9 Pentose phosphate pathway (p. 162)*
- Inosine monophosphate (IMP) is synthesized from PRPP via ten steps
- Purines AMP and GMP are synthesized in two steps from IMP (see Figure 7.3):
  - i. IMP → XMP → GMP
  - ii. IMP → adenylosuccinate → AMP
- All pyrimidines are derived from UMP (uridine monophosphate), which is itself synthesized in six steps from glutamine, aspartate, and bicarbonate
- CTP and dTMP are derived from UMP (see Figure 7.3):
  - i. UMP → UDP → UTP → CTP
  - ii. UMP → dUDP → dUMP → dTMP

Key evidence for synthetic pathways
- ^{14}C-labelled precursors used (e.g. $^{14}CO_2$; glycine: $NH_3^+CH_2{}^{14}COO^-$)
- The position of radiolabel in product reveals which residues were derived from which sources—helped to solve pathway (Buchanan and Greenberg)

Nucleotide salvage pathways

Synthesis of nucleotides *de novo* uses considerable amounts of energy and cannot be carried out by all tissues. Since many mRNA molecules are short-lived, the cell recovers nucleotides from degraded RNA (etc.) through salvage pathways. Products from DNA degradation can also be used for synthesis of nucleotides.

Nucleotides

Purines

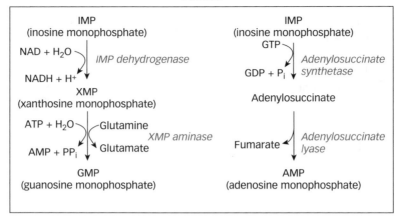

Pyrimidines

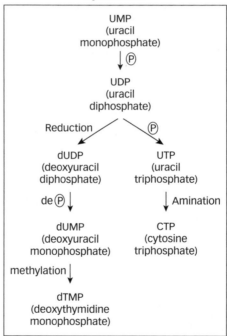

Figure 7.3 Nucleotide synthesis.

- The cell salvages nucleotides when they are present in the diet or in excess from cellular metabolism, e.g. short-lived mRNAs
- Intracellular nucleotides are more frequently salvaged than dietary nucleotides
- Nucleotide monophosphates can be salvaged from mononucleotides, nucleosides, and heterocyclic bases, using PRPP (the key molecule used for synthesis of Pu and Py) (see Figure 7.4)

- Deoxyribonucleotides can be formed from ribonucleotide diphosphates rNDPs (ADP, GDP, CDP) by _ribonucleoside diphosphate reductases_ (aka _ribonucleotide reductase_, or _RNR_) using reducing power of NADPH:
    i. dNTPs (except dTTP) then form from dNDPs by action of _nucleoside diphosphate kinases_
- dTTP formation:
    i. thymine is converted to dTMP by addition of deoxyribose-1-phosphate (_deoxythymidine phosphorylase_)
    ii. dTMP is then converted to dTDP and dTTP by _thymidylat kinase_
- rNTPs are substrates of _cobalamin-dependent reductases_ only in some micro-organisms
- UTP formation:
    i. uracil is converted to uridine monophosphate by addition of ribose-1-phosphate (_uridine phosphorylase_)
    ii. UMP is phosphorylated to form UDP and UTP by _uridylat kinase_
- Important enzyme activities in purine salvage include:
    i. _adenine phosphoribosyl transferase_ (_APRT_)
    ii. _hypoxanthine-guanine phosphoribosyl transferase_ (_HGPRT_)
- Pyrimidine salvage via _orotate phosphoribosyl transferase_ (_OPRT_)—an enzyme important also in _de novo_ synthesis
- Salvage pathways are essential—defects lead to human disease

Regulation of nucleotide synthesis

- Nucleotides are synthesized:
    i. throughout cell division cycle
    ii. _ribonucleotide reductase_ gene is transcribed prior to DNA synthesis (S phase) in the eukaryotic cell cycle: the enzyme is most active just prior to S phase in order to generate dNTPs from rNTPs to provide large nucleotide pools for synthesis of DNA

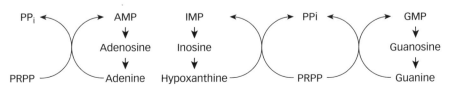

Figure 7.4 Nucleotide salvage pathway.

Nucleotide in activity site	Nucleotide in specificity site	Catalytic activity/specificity
dATP	–	blocked
ATP	ATP or dATP	CDP, UDP specific
ATP	dTTP	GDP specific
ATP	dGTP	ADP specific

Table 7.1 Regulation of ribonucleotide reductase (RNR)

- It is critical that purine and pyrimidine metabolism are linked in cells to balance the nucleotide pools as *DNA polymerases* are sensitive to nucleotide concentrations (because of complementary base pairing, Pu = Py in DNA) and are inhibited if amounts of purines and pyrimidines are unequal
- Regulation is via allosteric activation and inhibition of committed steps:
  - i. high concentrations of purines stimulate pyrimidine biosynthesis and block purine synthesis
  - ii. high concentrations of pyrimidines stimulate purine biosynthesis and block pyrimidine synthesis
- AMP inhibits the first step in its biosynthesis—feedback inhibition
- CTP synthesis is balanced with GTP synthesis: *CTP synthetase* is allosterically inhibited by CTP (pyrimidine) and activated by GTP (purine) (*E. coli*)
- Key enzymes in regulation:
  - i. *dihydrofolate reductase* (*DHFR*)
  - ii. *thymidylate synthase*
- Reduction by *ribonucleotide reductase* (*RNR*) is rate-limiting for dNTP formation
- One catalytic plus two allosteric regulatory sites on *RNR* ensure that excess of purine leads to synthesis of pyrimidine and vice versa (see Table 7.1):
  - i. activity site controls activity of the catalytic site
  - ii. specificity site controls substrate specificity of the catalytic site

Looking for extra marks?

Nucleotide metabolism is closely linked to cell division. Several anti-cancer treatments target nucleotide synthesis, e.g. methotrexate targets *dihydrofolate reductase* while 5-fluorouracil (5FU) is converted to 5-fluorodeoxyuridylate, which blocks *thymidylate synthase*.

7.2 STRUCTURAL LIPIDS

Key concepts: structural lipids

- Complex structural lipids such as membrane phospholipids, and steroids such as cholesterol, must be synthesized within the cell
- They can also be broken down to generate energy, though they do not constitute major catabolites under normal conditions

Breakdown of structural lipids

1. Phospholipids

- Breakdown of dietary phospholipids by pancreatic *phospholipase* (especially *PLA₂*) secreted into intestine
- *PLA₂* cleaves the C2 position of the phospholipid to yield fatty acid (arachidonic acid) and lysophospholipid (which acts as detergent at high concentrations— accounts for toxicity of snake, bee, and wasp venom which contain *PLA₂*)
- Lysophospholipid is absorbed by intestinal cells and then re-esterified to glycerophospholipid

2. Cholesterol

- Cholesterol esters taken up in the diet are hydrolysed by *esterase* in lumen of intestine
- Free cholesterol is solubilized in bile salt micelles for absorption
- Cholesterol is then esterified with acetyl CoA in intestinal cells

site of esterification on cholesterol

Looking for extra marks?

The membrane phospholipid, phosphatidyl inositol 4,5 bisphosphate (PIP_2), is broken down to IP_3 and diacylglycerol by intracellular *phospholipase C* (beta or gamma) that is activated by hormone binding to a specific membrane-bound receptor during cell signalling. PIP_2 can be formed by phosphorylation of phosphatidyl inositol.

➔ *see 6.5 Principles of hormonal signalling in metabolism (p. 183)*

PLA_2 is important in the inflammatory response, as PLA_2 generates arachidonic acid. This proinflammatory action is beneficial on transient infection or injury as it recruits immune cells, but detrimental in chronic inflammation e.g. in rheumatoid arthritis. Anti-inflammatory drugs such as aspirin indirectly block PLA_2 action.

Biosynthesis of lipids

Lipids, including triacylglycerols, phospholipids, sphingolipids, and steroids are all synthesized from a common intermediate, phosphatidate. Bacteria and mammals differ in the synthesis of certain phospholipids, and cholesterol synthesis is unique to animals. The majority of lipids are synthesized in the smooth endoplasmic reticulum in eukaryotes.

Structural lipids

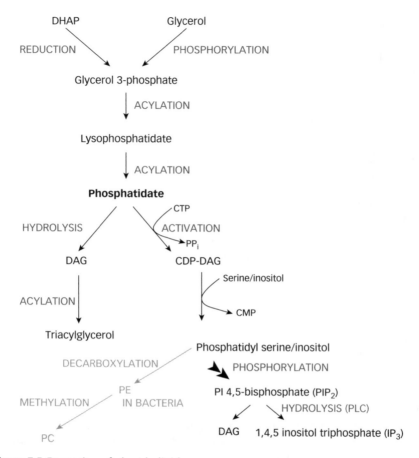

Figure 7.5 Formation of phospholipids.

1. Phospholipids

- Phospholipids originate from glycerol 3-phosphate, which is generated from glycerol or dihydroxyacetone phosphate (DHAP) (Figure 7.5)
- Phosphatidate, derived from glycerol 3-phosphate, is a common precursor of membrane phospholipids (PS, PI, PE, and PC), triacylglycerols, and signalling lipids
- Cytidine nucleotide (CTP) serves an activating role in formation of phospholipids—similar to role of uridine nucleotide in glycogen formation
- Diacylglycerol (DAG) is precursor of triacylglycerols (triglycerides)
- Mammalian PE is synthesized either from ethanolamine using ATP then CTP, or by exchanging the serine of PS for ethanolamine
- PC is synthesized in bacteria from PE by methylation, but from choline in the diet in mammals in three steps (see Figure 7.6)

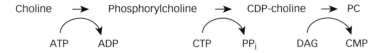

Figure 7.6 Synthesis of phosphatidyl choline in mammals.

2. Sphingolipids

- Backbone is sphingosine (see Figure 7.7) rather than glycerol for phospholipids
- Pyridoxal phosphate is required for enzyme activity to generate sphingosine
- Ceramide, formed by acylation of sphingosine, is precursor for cerebrosides and gangliosides (with sugars), or sphingomyelin, using CDP-choline as donor

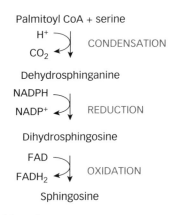

Figure 7.7 Formation of sphingosine.

3. Cholesterol

- Cholesterol (a 27-carbon ring structure with all carbons derived from acetyl CoA) serves important roles in animals:
    - i. modulates the fluidity of animal cell membranes (confers rigidity but also limits phase changes on shifts in temperature—prevents crystallization)
    - ii. precursor for formation of bile salts, steroid hormones, and vitamin D
- Mevalonate (C6), isopentyl pyrophosphate (C5), and squalene (C30) are intermediates in cholesterol synthesis
- Mevalonate synthesis = committed step in production of cholesterol
- Conversion of mevalonate to isopentyl pyrophosphate occurs in three steps, each using one molecule of ATP
- Squalene is formed from isopentyl pyrophosphate via C5 → C10 → C15 → C30
- Squalene cyclises to form lanosterol: requires molecular oxygen (may account for presence of cholesterol only in eukaryotes, i.e. evolved after atmosphere became aerobic)
- Reduction of lanosterol, removal of three methyl groups, and migration of remaining double bond results in synthesis of cholesterol

> ## Looking for extra marks?
>
> Formation of mevalonate, the key precursor also in cholesterol synthesis, is blocked by statins, so these drugs are used to reduce blood cholesterol in atherosclerosis.
>
> Note that many oils in plants (e.g. lemon, peppermint, geranium), also rubber and ubiquinone, are derived from isopentyl pyrophosphate.

7.3 GLYCOPROTEINS

> ## Key concepts: glycoproteins
>
> - Glycoproteins are formed by covalently linking sugars to proteins
> - Sugar residues are linked by an amide bond to the $-NH_2$ group of asparagine (N-linked), or the $-OH$ group of serine or threonine (O-linked)
> - Glycosylation occurs in the endoplasmic reticulum (ER) and in the Golgi body
> - Such sugar 'tags' can be important in directing proteins to the correct subcellular location (e.g. Golgi apparatus, lysosomes, plasma membrane), or for secretion from the cell

N-linked glycosylation

- Only asparagines in the sequences Asn-X-Ser or Asn-X-Thr are substrates for glycosylation
- Sugars are activated in the cytosol as sugar–nucleotide intermediates, which donate the sugar to the growing oligosaccharide
- Oligosaccharide composed of *N*-acetylglucosamine, mannose, and glucose is formed on a carrier lipid found in ER lumenal membrane—dolichol phosphate
- First seven sugars are added on cytosolic face of ER, then dolichol is 'flipped' across membrane, and remainder of sugars are added in lumen of the endoplasmic reticulum
- After transfer to protein, all three glucose residues and one mannose are 'trimmed'. Lack of glucose is a signal that the protein is ready for export to Golgi apparatus
- Further modifications and additions occur in Golgi depending on sequence of protein
- Two major types of N-linked oligosaccharides are generated:
    i. high mannose oligosaccharides have only extra mannose residues added in Golgi (e.g. lysosmal proteins carry a mannose-6-phosphate tag)
    ii. complex oligosaccharides have *N*-acetylglucosamine, galactose, sialic acid, and fucose residues added in varying amounts (e.g. cell adhesion molecules—CAMs)
- Modifications follow a highly ordered pathway as proteins pass sequentially through the various cisternae of the Golgi apparatus
- Glycoprotein chain length varies according to rate of passage through ER (faster passage → less glycosylation) resulting in heterogeneity (i.e. same protein can have different length carbohydrate chains)

O-linked glycosylation

- Formation is catalysed by a series of *glycosyl transferases* which use sugar-nucleotides as activated precursors
- Addition takes place in successive cisternae of Golgi apparatus
- *N*-acetylgalactosamine is added first
- Variable number of other sugar residues are then added
- e.g. nuclear pore complex proteins are modified by O-linked glycosylation (*N*-acetyl glucosamine, GlcNAc)
- e.g. chitin in insect exoskeletons is a polymer of GlcNAc

7.4 VITAMINS

Key concepts: vitamins

- **Vitamins** are essential components of the diet
- Complex organic molecules needed in trace amounts by the body
- Serve various critical roles in metabolism, often act as co-factors
- Dietary deficiency can lead to fatal disease
- Overdose of lipid-soluble vitamins can also result in disease
- See Table 7.2 (p. 200) for roles of vitamins in metabolism

Vitamin	Solubility	Functional product	Role in metabolism	Deficiency disease
A (retinol)	lipid	retinal	vision; necessary for growth of young animals	night blindness
B1 (thiamine)	water	thiamine pyrophosphate	transfer of aldehyde groups	beriberi (fatal disease, with pain, paralysis, atrophy, oedema, and cardiac failure)
B2 (riboflavin)	water	FMN (flavin mononucleotide) and FAD (flavin adenine dinucleotide)	redox reactions involving one or two electron transfers	rare—inflamed tongue, mouth lesions, and dermatitis
B3 (niacin)	water	NAD+ (nicotinamide adenine dinucleotide) and NADP+ (nicotinamide adenine dinucleotide phosphate)	redox reactions involving one or two electron transfers	pellagra (fatal disease, with diarrhoea, dermatitis, and dementia)
B5 (pantothenic acid, pantothenol)	water	Pantothenic acid	synthesis of coenzyme A (CoA)—transfer of acyl groups, and plays central role in metabolism	rare—impaired energy production, fatigue, neurological symptoms
B6 (pyridoxine)	water	pyridoxal 5' phosphate (PLP)	transfer of groups to and from amino acids; cofactor for glycogen phosphorylase in glycogen breakdown; lipid metabolism; neurotransmitter synthesis	dermatitis, atrophic glossitis, neuropathy
B7 (biotin)— previously known as vitamin H	water	biocitin	carboxyl group transfer or ATP-dependent carboxylation; co-enzyme in metabolism of fatty acids and leucine, role in gluconeogenesis	rare—dermatitis, conjunctivitis, alopecia; note that some egg allergies result from avidin in egg white binding to biotin (not a deficiency disease)
B9 (folic acid)	water	tetrahydrofolate	nucleotide synthesis: provides methyl group for thymine of DNA	megaloblastic anaemia; neural tube defects causing spina bifida if severe deficiency in pregnancy
B12 (cobalamin)	water	adenosylcobalamin methylcobalamin	intramolecular rearrangements; methyl group transfer (especially in formation of methionine); reduction of ribonucleotides to deoxyribonucleotides	pernicious anaemia (absorption requires intrinsic factor secreted in stomach—anaemia usually results from lack of intrinsic factor rather than lack of B12 in diet)
C (ascorbic acid)	water	ascorbate	hydroxylation of proline in stabilizing collagen structure	scurvy (fatal disease with skin lesions, fragile blood vessels, reduced wound healing)
D2 (ergocalciferol) D3 (cholecalciferol)	lipid	1,25-dihydroxychole-calciferol (a steroid hormone)	promotes absorption of dietary Ca^{2+}; synthesized in skin on exposure to UV light so light treatment of children necessary above Arctic Circle	rickets (stunted growth and deformed bones in children); overdose can also cause disease including bone demineralization, calcification of soft tissues, kidney stones, and kidney failure
E (tocopherol)	lipid	α-tocopherol	anti-oxidant: protects unsaturated membrane lipids from oxidation	ataxia and peripheral neuropathy, myopathy, impaired immunity, abortion, male sterility
K	lipid (synthetic forms can be water soluble)	menaquinone	carboxylation of some glutamate residues in Ca^{2+}-binding domains of some proteins; essential for the synthesis of prothrombin in blood clotting; role in bone metabolism	haemorrhage especially in newborn infants (hence given to neonates by injection)

8 Microbial and plant metabolism

Some important metabolic pathways are unique to microorganisms and plants. These include trapping of solar energy, and fixation of atmospheric nitrogen and carbon, so they are essential to (almost) all life.

Key concepts

- Light absorbed by chlorophyll molecules is used to drive **photochemistry** in **photosynthesis**
- Photons excite electrons, making them more reducing. The electrons can be used to reduce:
    i. electron transfer complexes, leading to the formation of a proton (H^+) gradient and hence ATP
    ii. cofactors such as NAD^+ and $NADP^+$, leading to the reduction of CO_2 and hence carbon fixation
- In plants and **cyanobacteria**, the 'excited' electrons are derived from H_2O with the evolution of O_2 as a by-product
- Primary acceptor for CO_2 in carbon fixation is ribulose-1,5-bisphosphate, catalysed by the enzyme *Rubisco*
- After the initial reaction with CO_2, sugars are interconverted in the **Calvin cycle** by reactions similar to those in the **pentose phosphate pathway**

continued

- O_2 and CO_2 compete for the active site of *Rubisco*. At ambient O_2 and CO_2 levels, CO_2 fixation is inefficient. Some plants (e.g. tropical grasses) have mechanisms for locally increasing CO_2 levels for *Rubisco*
- Plants also can carry out the phosphoglycolate pathway to reclaim the by-products of the reaction of ribulose-1,5-bisphosphate with O_2
- Nitrogen-fixing bacteria can reduce atmospheric N_2 to NH_3, providing nitrogen for the biosphere. These may be free-living bacteria, or associated with plant roots in nodules
- Enzyme involved in N_2 reduction is *nitrogenase*, which contains both Fe and Mo. $6e^-$ are required to reduce 1 mol N_2, and H_2 (from the reduction of H^+) is evolved as a by-product
- Most plants assimilate NH_3 by the reduction of NO_3^- from water in soil

8.1 PHOTOSYNTHESIS

Almost all the energy consumed by biological systems derives ultimately from solar energy trapped by photosynthesis carried out by green plants and photosynthetic bacteria.

- Photosynthetic bacteria trap light energy and use this for ATP synthesis
- Photosynthesis in plants and cyanobacteria also fixes carbon from CO_2 into carbohydrate, and evolves O_2 by splitting H_2O
- Photosynthetic evolution of O_2 by **autotrophs** has been essential for energy generation in **heterotrophic** organisms and allowed evolution of a wide range of life forms
- Major events of plant photosynthesis take place in the **thylakoid membrane** of **chloroplasts**
  ➔ *see 2.5 Organelles (p. 49) for chloroplast structure*
- Plant photosynthesis can be separated into two phases:
  i. light reactions (driven by photons)
  ii. dark (or light-independent) reactions driven by the chemical products of the light reactions (NADPH and ATP)

Light reactions of photosynthesis

A photon is used to excite a chlorophyll molecule. The excited chlorophyll is a much stronger reducing agent than in its ground state, and its excited electron is used to sequentially reduce other reducing agents and ultimately $NADP^+$ to NADPH. Energy released by flow of electrons through carriers is used to generate a proton gradient across the thylakoid membrane (similar to oxidative phosphorylation).

- Two interacting **photosystems** (PS) exist which absorb at slightly different wavelengths of light:
  i. PSI ≤ 700 nm
  ii. PSII ≤ 680 nm

- In each separate photosystem, photons absorbed by many **chlorophyll** molecules ('antennae chlorophyll') are funnelled into a reaction centre, where the photochemistry occurs
- Each photon at a reaction centre causes the expulsion of $1e^-$ from a closely apposed pair of chlorophyll molecules ('special pair') within PSI or PSII
- PSII contains a core of two related proteins (the reaction centre dimer), which catalyses the light-driven reduction of plastoquinone
- PSII$^+$ (after the electron is removed) can 'pull' electrons from water to generate oxygen, using a 4Mn centre
- Electrons flow from PSII to PSI via cytochrome bf and plastocyanin (PC), and the energy is used to generate H$^+$ gradient (see Figure 8.2)
- PSI also contains a core of two related proteins (larger than those in PSII) which catalyse the light-driven transfer of these electrons from plastocyanine to NADP$^+$ $\rightarrow$ NADPH
- The proton gradient is used to drive ATP synthesis (**photophosphorylation**) via a proton-translocating *ATP synthase* (CF$_1$F$_o$) in the thylakoid membrane, very similar to that found in the inner mitochondrial membrane in oxidative phosphorylation. Approximately one ATP molecule is produced per pair of photons absorbed by PSII
- The redox span between H$_2$O and NADPH is approximately 1.1 V (i.e. the electrons are driven 'uphill') (see Figure 8.1)
- Cooperation of the two photosystems in exciting electrons can be represented in the Z-scheme (Figure 8.1)
- Each photosystem moves electrons from the luminal side of the thylakoid to the stromal side (Figure 8.2)
- Note similarity of photosynthetic electron transport chain with that used in oxidative phosphorylation

  ⮞ *see 5.4 Oxidative phosphorylation (p. 140) and Figure 5.5 (p. 141)*

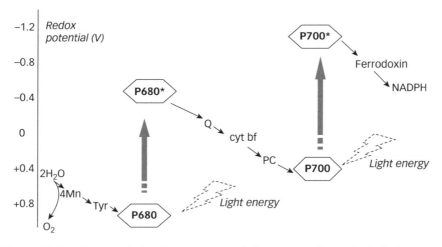

Figure 8.1 The 'Z-scheme', showing energetics of electron transfer in plant photosynthesis. Note redox scale on vertical axis; upwards = more reducing.

Photosynthesis

- Alternatively, PSI can generate a cyclical flow of electrons to generate ATP without the reduction of NADP+

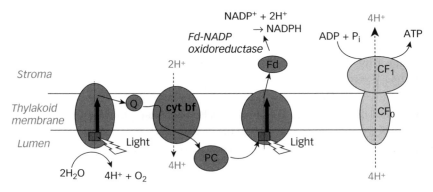

Figure 8.2 Membrane organisation in photosynthetic light reactions. Electron transfer is shown as solid arrows, H+ transfer as dotted arrows. Stoichiometry of H+ transfer relates to 1 QH$_2$ oxidized (cyt bf) or 1 ATP made (ATP synthase).

Looking for extra marks?

Many **herbicides** (e.g. dichloromethylurea) inhibit photosynthesis, so are specific for plants and do not affect animals.

Light-independent reactions of photosynthesis

This series of reactions uses reducing power and ATP generated in light reactions to assimilate CO_2 into organic compounds, but does not directly use photons. Ribulose-1,5-bisphosphate is the initial acceptor for CO_2 and is regenerated in subsequent reactions, with glyceraldehyde phosphate the primary product.

CO_2 fixation

- One CO_2 molecule condenses with a five-carbon sugar, ribulose-1,5-bisphosphate (RuBP), to generate an intermediate compound with six carbon atoms (Figure 8.3)
- This carboxylated intermediate is rapidly hydrolysed to generate two molecules of 3-phosphoglycerate (3C units)
- Catalysed by *ribulose-1,5-bisphosphate carboxylase/oxygenase* (*Rubisco*—see below)
- 3-phosphoglycerate is then reduced to glyceraldehyde 3-phosphate (GAP) by NADPH
- ATP is also required as the reaction proceeds via the intermediate 1,3-diphosphoglycerate
- RuBP is regenerated by a series of reactions known as the **Calvin cycle** with a net gain of one GAP per three CO_2 fixed
- Three ATP and two NADPH are used per molecule of CO_2 fixed

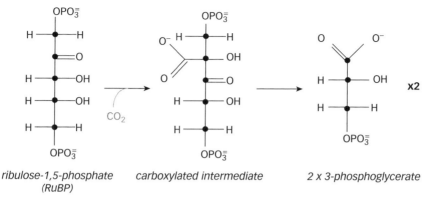

| ribulose-1,5-phosphate (RuBP) | carboxylated intermediate | 2 x 3-phosphoglycerate |

Figure 8.3 Carboxylation by Rubisco. A black dot represents a C atom.

Rubisco

- *Ribulose-1,5-bisphosphate carboxylase-oxygenase* (*Rubisco*) is the enzyme responsible for light-dependent fixation of CO_2 into 3C compounds in plants and photosynthetic bacteria
- Constitutes ~50% of soluble protein in plant leaves
- Probably the most abundant protein on earth
- Stromal concentration of *Rubisco* active sites ~4 mM
- Low **turnover number**
- Stringently regulated:
    i.   inhibited by ribulose-1,5-bisphosphate
    ii. activity requires CO_2, Mg^{2+} (3–6 mM), and correct stromal pH (around 8)
    iii. some plants synthesize the inhibitor carboxyarabinitol-1-phosphate.
- Active in light, as stromal pH and $[Mg^{2+}]$ increase
- *Light- and ATP-dependent Rubisco activase* converts *Rubisco* to a conformation that releases active site inhibitor
- Activating (non-substrate) CO_2 molecule forms carbamate with lysine side chain of enzyme. *Rubisco* is inactive in dark—carbamate dissociates
- Higher plant *Rubisco* is composed of eight large subunits (chloroplast genome-encoded) and eight small subunits (encoded by nuclear genome); cf. bacterial *Rubisco* has only eight large subunits
- *Rubisco* can also catalyse the addition of oxygen to RuBP in a metabolically wasteful side reaction—**photorespiration**
  ➔ *see 8.2 Ancillary reactions of photosynthesis (p. 208)*

Looking for extra marks?

Remember that *Rubisco* is probably the most abundant protein on earth.

Photosynthesis

Calvin cycle (reductive pentose phosphate pathway)

- Follows on from reductive phase (above):

$$3 \text{ RuBP} + 3 \text{ CO}_2 + 6\text{ATP} + 6 \text{ NADPH} \rightarrow$$
$$6\text{GAP} + 6 \text{ NADP}^+ + 6 \text{ ADP} + 6 \text{ P}_i$$

- One GAP used for biosynthesis, and five remaining GAP re-form three molecules of ribulose 5-phosphate (Ru5P)—no further expenditure of NADPH or ATP
  - ➡ *see 5.9 Pentose phosphate pathway (p.162)*
- R5P is then phosphorylated using ATP to generate RuBP by *phosphoribulokinase*
- Overall stoichiometry: $5 \text{ C}_3 \rightarrow 3 \text{ C}_5$
- Note that neither ATP nor glucose can be exported from chloroplasts; exported compound is GAP

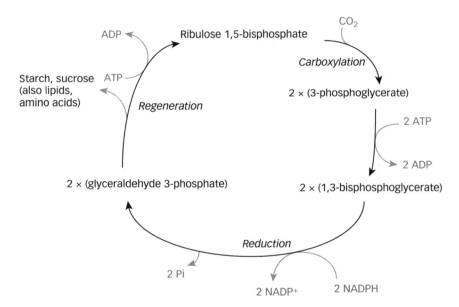

Figure 8.4 The Calvin cycle.

Regulation of the Calvin cycle

- In daylight, the two reactions of photosynthesis (light and dark) prevail and carbohydrate is synthesized
- At night, catabolism of carbohydrate generated during the day is used to generate energy
- Three mechanisms of control:
  - i. stromal pH changes

ii. [Mg^{2+}] changes

iii. redox changes

- In light, electron transfer pumps H$^+$ into chloroplast lumen and stromal pH rises—*Rubisco* and other enzymes (e.g. *FBPase*) are activated by high pH
- H$^+$ moving into chloroplast lumen expels Mg^{2+} and [Mg^{2+}] in stroma rises. This activates *Rubisco*
- Light reactions generate NADPH, so stroma becomes more reduced—sensed initially by **thioredoxin**, a small protein which contains an S–S bridge that can exist in reduced (2SH) or oxidized S–S forms. Reducing conditions bias equilibrium towards –SH
- Reduced thioredoxin reduces S–S bridges in carbohydrate biosynthetic enzymes (e.g. *Ru5P kinase, FBPase*) switching them on, and in carbohydrate-degradative enzymes (e.g. *glucose 6-phosphate dehydrogenase*) switching them off
- Hence biosynthesis and catabolism of carbohydrates are coordinated by thioredoxin

Evidence for mechanism of photosynthesis

- Plants can replenish oxygen in air which has been depleted by experimental combustion
- Tracing heavy oxygen (^{18}O) revealed that O$_2$ released came from H$_2$O and not CO$_2$
- Isolated chloroplast membranes can release oxygen in response to light in the absence of CO$_2$, i.e. separate light and dark reactions
- Photosynthetic efficiency increases synergistically when simultaneously stimulated with light of two different wavelengths (680 nm + 700 nm)—demonstrates cooperation of two PS reaction centres with different absorption wavelengths
- Chlorophyll fluorescence transitions indicate changes in distribution of the antennae chlorophylls between photosystems. (Fluorescence indicates inefficient light trapping, as the photon is re-emitted rather than used to drive photochemistry.)
- Artificial generation of a pH gradient across the thylakoid membrane can drive ATP synthesis in the absence of electron transfer
- Use of radioactively labelled CO$_2$ (^{14}C) allowed the path of carbon fixation to be deduced

 Check your understanding

Short answer questions

Draw a chloroplast. Indicate on your diagram the locations of: (a) the photosystems; (b) *Rubisco*; (c) the transmembrane pH gradient (give direction). (*Hint: you should combine information from Figure 2.11 and Figure 8.2.*)

continued

How is *Rubisco* regulated? (*Hint: all the information you need is in the section above on Rubisco.*)

How do photons of light generate ATP? (*Hint: remember that PSI alone can generate a pH gradient.*)

8.2 ANCILLARY REACTIONS OF PHOTOSYNTHESIS

Photorespiration

This apparently wasteful pathway is due to the oxygenase activity of *Rubisco* in which ribulose 1,5-bisphosphate is oxygenated to form 3C 3-phosphoglycerate and 2C phosphoglycolate. Phosphoglycolate is of no use to the plant, and its carbon skeleton must be salvaged, although 1 mol of CO_2 is lost (per 2 mol phosphoglycolate) in this process.

- Oxygenation of ribulose-1,5-bisphosphate catalysed by *oxygenase* activity of *Rubisco*:
  i. *oxygenase* activity is at same active site as *carboxylase* activity—both activities compete
  ii. despite a strong preference for CO_2, *oxygenase* activity of Rubisco is 0.25–0.33 times that of *carboxylase* activity at 25°C and atmospheric conditions due to the higher concentration of O_2 in air (stromal $[CO_2] = 10$ μM, $[O_2] = 250$ μM)
  iii. *oxygenase* activity increases with increased temperature and light intensity, i.e. tropical conditions
- Products are 3-phosphoglycerate and 2-phosphoglycolate (see Figure 8.5):
  i. 3-phosphoglycerate enters Calvin cycle
  ii. Phosphoglycolate is salvaged (Figure 8.6) by conversion to 2C amino acid, glycine, and the combination of two glycines to form the 3C amino acid, serine, with loss of CO_2
- Net effect is light-dependent uptake of O_2 with release of CO_2 (hence photorespiration)
- Photorespiration can greatly decrease crop yields as plants release CO_2 rather than fix it

Looking for extra marks?

Photorespiration has remained over millions of years of evolution, and no known mutations of Rubisco remove oxygenase activity, suggesting a possible essential physiological role for photorespiration. There are several hypotheses to explain this:

- Photorespiration generates ADP and $NADP^+$ when $[CO_2]$ is low (stomata closed, high intensity light)—that may prevent accumulation of an excess proton gradient (and hence membrane damage) across the thylakoid membrane

continued

- Photorespiration increases oxygen consumption, which may protect pigments and membrane lipids from damage by reactive oxygen species formed by the photosystems

C4 (Hatch–Slack) pathway

Some plants avoid wasteful photorespiration by *spatially* separating carbon fixation, in mesophyll cells, from carbon utilization, in bundle sheath cells (Figure 8.7). The initial CO_2 acceptor is phosphoenol pyruvate (PEP). Effectively this constitutes a CO_2 pump to the site of *Rubisco*.

- Species include tropical crops such as maize, sorghum, and sugarcane, plus many weeds
- CO_2 fixation occurs in mesophyll cells, near exterior of leaf, while Calvin cycle occurs in interior bundle sheath cells which have lower $[O_2]$ than mesophyll
- Mesophyll cells of C4 plants do not contain *Rubisco*:
  i. HCO_3^- is fixed by *PEP carboxylase*, avoiding CO_2/O_2 competition
  ii. initial product of carbon fixation is 4C acid (oxaloacetate) rather than 3C
  iii. carbon is transported to bundle sheath cells as malate or aspartate
  iv. CO_2 released by oxidative decarboxylation of malate in bundle sheath cells which contain *Rubisco*
  v. 3C acceptor transported back to mesophyll as pyruvate
- Five ATP and two NADPH used per mol CO_2 fixed (cf. 3ATP/2NADPH in C3 plants)

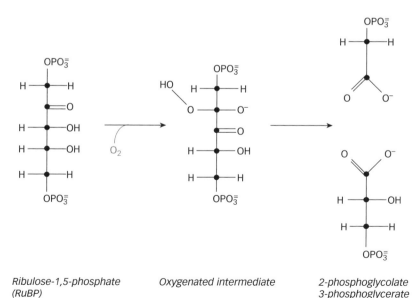

| Ribulose-1,5-phosphate (RuBP) | Oxygenated intermediate | 2-phosphoglycolate 3-phosphoglycerate |

Figure 8.5 Photorespiration catalysed by Rubisco. A black dot represents a C atom.

Ancillary reactions of photosynthesis

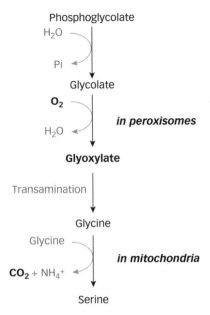

Figure 8.6 Salvage of phosphoglycolate during photorespiration.

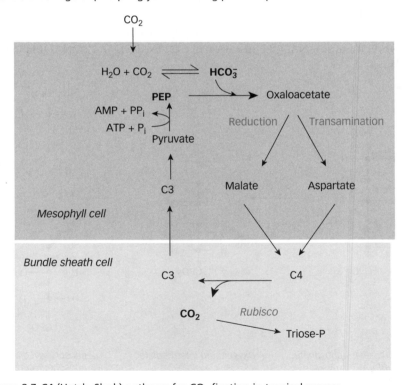

Figure 8.7 C4 (Hatch–Slack) pathway for CO_2 fixation in tropical grasses.

- Additional ATP used for 'transport' and formation of PEP from pyruvate
- C4 pathway is costly but gives advantage (high biomass production) over C3 plants in tropical conditions (high light intensity, high temperatures) due to lack of photorespiration

Key evidence

- Pulse with $^{14}CO_2$: ^{14}C appears initially in 4C acids, e.g. malate and aspartate
- Presence of key enzymes: *PEP carboxylase* and *pyruvate dikinase*

Crassulacean acid metabolism (CAM) pathway

The CAM pathway is an alternative method for avoiding photorespiration by *temporal* separation of carbon fixation (night) from carbon use (day). CO_2 is temporarily fixed during the night and stored as malate in the vacuole, and then released to *Rubisco* during daylight (Figure 8.8).

- Found in succulent plants in arid conditions, e.g. cacti, orchids, and bromeliads
- Temporal separation of CO_2 fixation from Calvin cycle
- Decreases water loss by temporarily fixing CO_2 at night (cooler) when stomata open

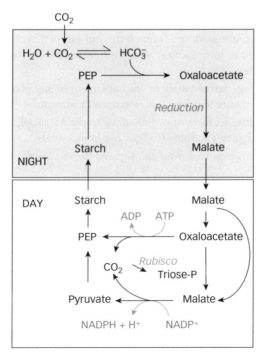

Figure 8.8 Crassulacean acid metabolism (CAM) pathway for CO_2 fixation in desert succulents.

Ancillary reactions of photosynthesis

	Night	*Day*
Stomata	open	closed
CO_2 handling	HCO_3^- from air fixed by *PEP carboxylase* reaction to form oxaloacetate	CO_2 generated by decarboxylation of malate fixed by *Rubisco*
Malate formation	oxaloacetate reduced to malate	
Fate of malate	malate is stored in vacuole (up to 0.2 M, vacuole occupies ≥90% of cell volume)	malate is released from vacuole and decarboxylated to pyruvate, which is converted to phosphoenolpyruvate (PEP)
Starch synthesis/breakdown	PEP is derived from starch stored in the chloroplast via glycolysis	PEP is formed into starch via gluconeogenesis and stored in chloroplast
Energy	ATP from glycolysis and oxidative phosphorylation	ATP is formed by light reaction of photosynthesis
Reducing power	NADH reduces oxaloacetate to malate	NADPH is formed by light reaction of photosynthesis and *malic enzyme*

Table 8.1 Comparison of light and dark photosynthesis reactions in CAM plants

- *PEP carboxylase* reacts HCO_3^- with PEP; *malic enzyme* splits malate and releases CO_2
- Tightly regulated: *PEP carboxylase* is inhibited by dephosphorylation, induced by malate and low pH, which prevents futile cycling of CO_2 and malate

Starch and sucrose conversions

Plants store carbohydrate produced from photosynthesis within chloroplasts as starch. This can be mobilized to form sucrose, which is transported around the plant to non-photosynthetic tissues.

- **Starch** is the storage carbohydrate in chloroplasts of higher plants, and in storage plastids in other tissues (e.g. root) (cf. glycogen in mammals)
- Sucrose is the transport form of carbohydrate in plants, supplying non-photosynthetic tissues, e.g. roots (c.f. glucose in mammals)
- Triose phosphate is exported from chloroplasts for sucrose synthesis (*not* glucose or ATP)
- Starch contains long polymers of α-1,4-linked glucose units, with a few α-1,6 branches

- Sucrose is a disaccharide of α-glucose 1,2 linked to β-fructose
- Cellulose is the carbohydrate component of plant cell walls, with β-1,4 linkages

- Starch in mammalian diet is broken down to glucose by action of *α-amylase* (secreted by salivary glands and pancreas)
- Cellulose cannot be broken down (digested) by mammals
- Yeast and bacteria instead store dextran polysaccharide (almost exclusively α-1,6 glycosidic linkages with occasional α-1,2, α-1,3, and α-1,4 linkages, according to species)

Starch and sucrose synthesis

- Glyceraldehyde 3-phosphate is generated inside chloroplasts in light (Calvin cycle)
- Two triose phosphates are condensed to form glucose phosphates (Figure 8.9)
- Glucose 1-phosphate is activated as ADP-glucose (ADPG) (for starch synthesis) or UDPG (for sucrose synthesis)
  - ➔ *compare with glycogenesis; see 5.5 Glycogen breakdown and synthesis (p. 145)*
- *Starch synthetase* extends primer using ADPG as donor
- *Branching enzyme* transfers short region from reducing end to a position a few residues away from new end, forming α-1,6 linkage
- Starch stored temporarily in chloroplasts is broken down to glucose-1-phosphate by *starch phosphorylase* at night
  - ➔ *compare with glycogenolysis; see 5.5 Glycogen breakdown and synthesis (p. 144) on glycogenolysis*
- Glucose-1-phosphate can then enter glycolysis to form triose phosphate which is exported from the chloroplast
  - ➔ *see 5.2 Glucose breakdown and synthesis (p.131) for glycolysis*
- Triose phosphates are converted to G1P (and hence UDPG) and F6P for sucrose synthesis in cytoplasm

Looking for extra marks?

Fructose 2,6-bisphosphate regulates the balance between starch and sucrose synthesis.

Ancillary reactions of photosynthesis

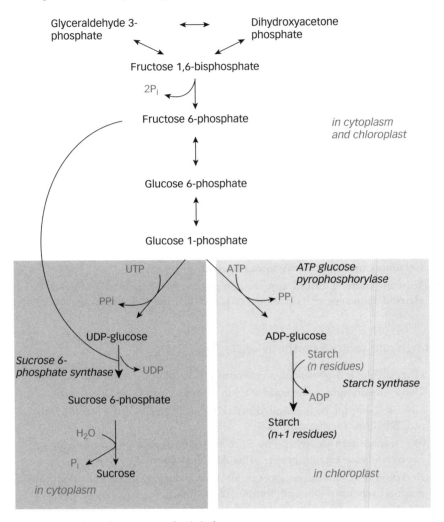

Figure 8.9 Starch and sucrose synthesis in leaves.

Glyoxylate cycle

Plants need to make carbohydrates from fatty acids since many seeds store energy as fats/oils. The glyoxylate cycle allows plants and bacteria to use acetyl CoA skeletons from fatty acid breakdown for de novo production of glucose.

- Glyoxylate cycle is effectively a shunt within the TCA cycle (Figure 8.10):
  - ➔ *see 5.3 Tricarboxylic acid (TCA) cycle (p. 137)*
    - i. bypasses the two decarboxylation steps of the TCA cycle
    - ii. two acetyl CoA moleculer enter each cycle (cf. only one in TCA cycle)

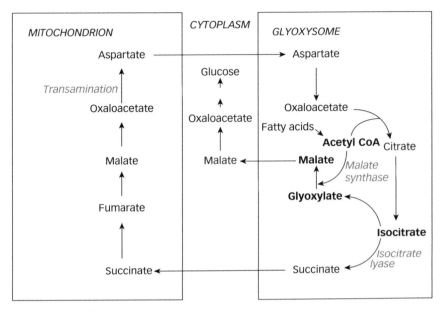

Figure 8.10 The glyoxylate cycle in plants.

- In plants, different parts of the pathway take place in mitochondria, cytosol, and **glyoxysomes** (specialized plant peroxisomes)
- Novel steps involve:
  - i. breakdown of isocitrate to succinate and glyoxylate (see Figure 8.10)
  - ii. condensation of this glyoxylate with acetyl CoA to form malate
- Malate is then exported into cytoplasm for gluconeogenesis
- Animals cannot use acetyl CoA to synthesize glucose (although fatty acid synthesis from carbohydrate is common). However, symbiotic bacteria in the gut of some ruminants can carry out glyoxylate cycle

 Check your understanding

How can plants generate glucose from fatty acids whereas animals cannot? (*Hint: glyoxylate cycle.*)

How do plants avoid photorespiration? (*Hint: remember CAM and C4 pathways.*)

8.3 NITROGEN ASSIMILATION

All organisms need NH_3 as a source of nitrogen for amino acid and nucleotide synthesis. Some strains of bacteria can reduce atmospheric N_2 to NH_3. Plants and autotrophic bacteria rely on inorganic sources of nitrogen, reducing NO_3^- to NH_3. (Animals obtain nitrogen from proteins in the diet.)

Ammonia production from N_2 (nitrogen fixation)

Nitrogen fixation is carried out by some bacteria and blue-green algae (cyanobacteria) which catalyse reduction of N_2 to ammonia (NH_3). This is a thermodynamically favourable process but, because of difficulty in breaking the triple bond in N_2, lots of ATP is used.

- Reduction of N_2 is thermodynamically favourable ($\Delta G^{0\prime} = -33$ kJ/mol), but N_2 is difficult to reduce kinetically
- Each N_2 reduced requires $8e^-$ and the hydrolysis of 16 ATP molecules:

$$N_2 + 8e^- + 8H^+ + 16ATP + 16H_2O \rightarrow 2NH_3 + H_2 + 16ADP + 16\,P_i$$

- H_2 is produced as a by-product by the reduction of H^+ ions in water
- N_2H_2 and N_2H_4 ($2e^-$ and $4e^-$ reduction products) appear to be intermediates on the pathway
- Electrons are donated from reduced **ferrodoxin** which is generated during photosynthesis, or by oxidative processes
- ATP assists the process by increasing the reducing power of the electrons (reducing $E^{0\prime}$)
- *Nitrogenase* complex consists of two proteins:
    i. *dinitrogenase reductase*, which provides electrons with high reducing power (Fe protein)
    ii. *dinitrogenase* carries out the reduction (Fe–Mo protein)
- Both *reductase* and *dinitrogenase* contain iron–sulphur clusters that are essential for electron transfer (Figure 8.11):
    i. *reductase* contains two ATP binding sites per Fe/S centre (hence 2ATP per electron above)
    ii. ATP binding distorts the Fe/S centres giving the electrons a greater 'push' towards the *dinitrogenase* metal centres
    iii. *dinitrogenase* contains Fe/S centre and Mo/Fe/S cluster (FeMoCo) where N_2 (probably) binds. The FeMoCo binds a homocitrate molecule (function unclear)
    iv. arrangement of these metal sulphur centres makes them very susceptible to O_2
- NH_4^+ generated is then assimilated into amino acids via *glutamate dehydrogenase*
  (➔) *see 5.8 Amino acid breakdown and synthesis (p. 160) for amino acid synthesis*
- Plants cannot themselves fix N_2, but some (e.g. legumes) may have nitrogen-fixing bacteria in root nodules (*Rhizobium* spp.) and take up nitrogen from symbiont

Looking for extra marks?

The metal centres in **nitrogenase** are very susceptible to O_2 and normally protected, e.g. by anaerobic growth, O_2 binding proteins (leghaemoglobin), or specialized cells (heterocysts).

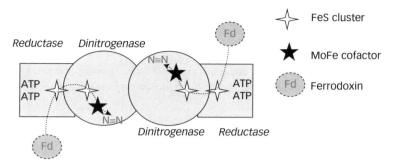

Figure 8.11 Organisation of *nitrogenase*.

Ammonia production from nitrate

Plants (except carnivorous ones) use soil nitrate as a source of NH_3. NO_3^- is reduced initially to nitrite, NO_2^- (2e⁻) which is subsequently reduced to NH_3 in a 6e⁻ reduction.

- Reduction of nitrate:

$$NADH + H^+ + NO_3^- \rightarrow NAD^+ + H_2O + NO_2^-$$

- Path of electrons is NADH→FAD→heme b→molybdopterin→NO_3^-, with three cofactors present on same polypeptide
- Reduction of nitrite:

$$6Fd^- + NO_2^- + 7H^+ \rightarrow 6Fd + NH_3 + 2H_2O \text{ (Fd}^- = \text{reduced ferrodoxin)}$$

- Path of electrons is Fd⁻→(Fe_4S_4) centre→siroheme→NO_2^- with two cofactors present on same polypeptide
- Partially reduced intermediates (NO, NH_2OH) remain bound to siroheme Fe during the 6e⁻ process
- NH_4^+ generated is then assimilated into amino acids via *glutamate dehydrogenase*
  ➜ *see 5.8 Amino acid breakdown and synthesis (p. 160) for amino acid synthesis*
- Although NO_3^- is taken up by the roots, nitrate reduction occurs largely in the shoots/leaves using the reducing power generated in photosynthesis
- Plant pathway, accumulating NH_3 in proteins and nucleotides, is 'assimilatory' nitrate reduction
- Some bacteria reduce nitrate to N_2, generating energy (effectively NO_3^- replaces oxygen as a terminal electron acceptor). This pathway, which loses N from the biosphere, is 'dissimilatory' and is carried out by denitrifying bacteria, using *nitrate reductase* and *nitrite reductase* unrelated to the plant enzymes

Check your understanding

Short answer questions

How is ammonia generated in biological systems? (*Hint: remember it can be generated from nitrogen OR from nitrate; think also about formation of ammonia from amino acid breakdown.*)

How do plants assimilate nitrogen? (*Hint: remember that plants use NO_3^- that is initially fixed by nitrogen-fixing bacteria.*)

Essay questions

Compare and contrast ATP generation by photosynthesis and oxidative phosphorylation. (*Hint: you will need to integrate material from several parts of your course. Look in particular at 8.1 Photosynthesis and 5.4 Oxidative phosphorylation for information on both processes.*)

What reactions are catalysed by *Rubisco* and how are they regulated? (*Hint: remember photorespiration as well as photosynthesis.*)

8.4 ETHANOL PRODUCTION

Ethanol is a highly reduced, two-carbon compound produced from glucose by some bacteria or fungi under anaerobic conditions. It is a commercially important product of fermentation by brewers' yeast, *Saccharomyces cerevisiae.*

- Ethanol is produced by the anaerobic breakdown (**fermentation**) of glucose by:
    i.   some bacteria (including gut bacteria in mammals)
    ii. fungi (e.g. *Saccharomyces cerevisiae*—brewers' yeast)
- Glycolysis yields pyruvate + NADH
- In anaerobiosis, NADH cannot be oxidized by O_2 so pyruvate undergoes decarboxylation and reduction, producing ethanol as a final product (ethanolic fermentation):
    i.   decarboxylation of pyruvate is catalysed by *pyruvate decarboxylase*, using the cofactor (B-vitamin) thiamine pyrophosphate
    ii. reduction of ethanal by *alcohol dehydrogenase* uses NADH generated in the glycolytic pathway, allowing glycolysis to continue in the absence of oxygen

➔ *see 5.10 Lactate and ethanol metabolism (p. 164) for ethanol breakdown in animals*

pyruvate   ethanal   ethanol

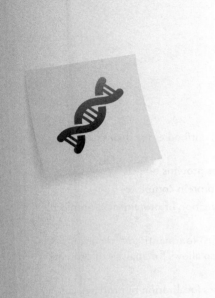

9 Biochemical techniques

The experimental procedures by which biochemical information is obtained including molecular biological, immunological, biochemical, and biophysical techniques.

Key concepts

Molecular biology techniques

These involve the analysis and manipulation of DNA, RNA, and protein. They include:

- Cloning genes/cDNAs
- Genomics: determining the sequence of DNA or whole genomes (e.g. human genome project)
- Transcriptomics: assessing gene expression
- Proteomics: assessing protein expression, also expressing and analysing recombinant proteins
- Gene therapy and genetically modified (GM) crops

Biochemical techniques

These allow separation of cell components (especially proteins) according to their biochemical properties and include:

- Affinity tag purification

continued

219

- Ion exchange chromatography
- Size exclusion gel filtration chromatography
- Analytical gradient centrifugation

Immunological techniques

These exploit the high specificity and affinity of antibodies for their cognate antigens, and include:

- Immunoblotting (aka western blotting)—for proteins on gels
- Immunoprecipation to isolate proteins or protein complexes
- Radioimmunoassay for quantitative determination of protein/macromolecule concentration
- Enzyme-linked immunosorbent assay (ELISA) for quantitative determination of protein/macromolecule concentration, also allows for analysis of protein–protein interactions
- Immunofluorescence microscopy—to assesses localization of proteins, e.g. within a cell, and under some conditions, can examine protein–protein interactions

Biophysical techniques

These allow analysis of the structure and physical properties of biochemical macromolecules (especially proteins and nucleic acids, also lipids):

- X-ray crystallography—for 'static' images of macromolecules that crystallize
- Single particle electron microscopy (EM) and cryoelectron microscopy
- Nuclear magnetic resonance spectroscopy (NMR)—allows determination of structure in solution of proteins <50 kDa
- Mass spectrometry (MS)
- Electron spin resonance (ESR)
- Surface plasmon resonance (SPR)

Note that other cell biological and genetic techniques are covered in the accompanying text books, *Thrive in Genetics* and *Thrive in Cell Biology*.

9.1 DNA MANIPULATION AND ANALYSIS

DNA concentration determination

- Absorption of ultraviolet (UV) light at wavelength of 260 nm A_{260} is determined
  i. DNA concentration is given by: $50 \times A_{260} = $ [DNA] in µg/ml
  ii. e.g. an optical density (OD) of 1.0 at 260 nm indicates a DNA concentration of 50 µg/ml
- Purity is assessed by also measuring absorption at 230 and 280 nm (peak absorbance wavelengths for carbohydrates and proteins, respectively). A ratio of 1:1.8:1 at $A_{230}:A_{260}:A_{280}$ indicates the DNA is pure

Restriction enzymes

Restriction enzymes cleave double-stranded DNA (dsDNA) at a specific sequence, and are used in molecular biology to cut DNA prior to further manipulation such as cloning, or to identify DNA by a unique pattern of restriction fragments (DNA fingerprinting).

- Produced by bacteria as 'immune system' to degrade unrelated foreign DNA (e.g. bacteriophage) that gets into bacterial cells

- Recognition sequence is usually a palindrome of four or six base pairs (bp) on dsDNA (**restriction site**) so can be used experimentally to cleave DNA at known sites

- Bacteria often methylate their own DNA on specific adenine residues within *restriction enzyme* recognition sites, but foreign DNA will not be methylated so it can be cleaved (see Figure 9.1)

- Cut both DNA strands to produce an overhanging (5′ or 3′) 'sticky' end or a blunt end (i.e. no overhang)

- Enzymes that recognise six bases will cut on average every 4096 bp (4^6) in DNA; those that recognize four will cut on average every 256 bp (4^4)

- Restriction map shows positions of enzyme cleavage sites on a specific piece of DNA

- Can be used to confirm identity of purified plasmid DNA for which the restriction map is known—cleavage products separated on agarose gels give characteristic patterns

- Used extensively during cloning to cut out a required DNA segment—this fragment can then be purified according to its size on an agarose gel

- Compatible DNA ends are created when DNA from two different sources (e.g. genomic DNA and plasmid vector) is cut with the same enzyme, facilitating their ligation

- Modern cloning vectors contain multiple restriction enzyme sites in a region known as a polylinker or MCS (multicloning site)

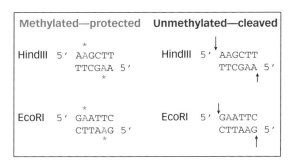

Figure 9.1 Examples of restriction enzyme sites. Note that cleavage can often be prevented by methylation of A—this is a mechanism to prevent bacteria from digesting their own DNA.

DNA manipulation and analysis

DNA ligase

DNA ligase seals nicks in the backbone of DNA by reforming phosphodiester bonds.

- Physiological role in DNA replication and repair
- Used in cloning to covalently join different DNA fragments, e.g. genomic DNA fragment and plasmid vector
- Precisely joins together two ends of DNA with complementary overhanging sequences generated by *restriction enzymes*
- Some bacterial and phage DNA ligases can seal blunt ends
- Uses ATP and requires a 5′ phosphate group on the DNA to be joined (two-step reaction):
  - i. can block self-ligation of a vector cut with a *restriction enzyme* by removing the phosphate groups (treat with *phosphatase*)
  - ii. still allows ligation of foreign DNA into vector as the foreign insert provides the 5′ phosphates necessary for ligation

DNA gels

- DNA can be separated according to size on gels:
  - i. polyacrylamide gels for small DNA fragments (single nucleotides to a few 100s). Can resolve fragments that differ by as little as a single nucleotide in length, so used to analyse products of DNA sequencing reactions
  - ii. agarose gels (for fragments 100 nucleotides to ~50 kb)
  - iii. pulsed field agarose gels for whole chromosomes (megabases)
- DNA is visualized on gels using intercalating dyes, e.g. ethidium bromide or SYBR Green under UV or blue light excitation
- Gels are used to determine size, purity, and approximate amount of DNA

Southern blots

A method for determining the presence and copy number of a DNA molecule of which at least some of the sequence is known.

- DNA is fragmented using restriction enzymes and separated on an agarose gel—if fragments are very large (>15kb), the gel may be exposed to acid to break the DNA within individual gel bands into smaller pieces to increase efficiency of transfer
- DNA is transferred to a nitrocellulose/nylon membrane by capillary action in alkali buffer to denature the DNA, or in neutral buffer
- The membrane is baked or UV cross-linked to fix the DNA onto the membrane
- Single-stranded radiolabelled DNA **probe** complementary to the DNA on the blot is incubated with the membrane
- DNA can be radioactively labelled on either the base (e.g. ^{3}H-thymidine) or the phosphate (e.g. α^{32}P-dATP)
- Unbound probe is washed off
- The blot is exposed to X-ray film and a band is observed where the probe has bound to the blot (see Figure 9.2)

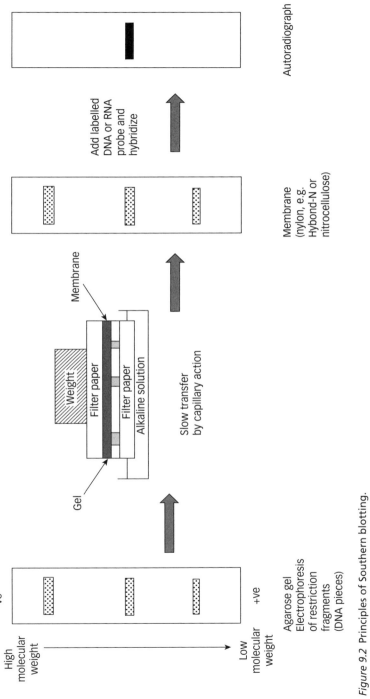

Figure 9.2 Principles of Southern blotting.

Different levels of stringency can be achieved by altering temperature, salt concentration, and detergent, affecting how strongly the probe can bind to the DNA on the blot, i.e. how closely matched the probe must be to the DNA on the gel. This is useful, for example, when using a probe designed against a known gene in one species to look for a homologous but non-identical gene in another species ('zoo blots').

DNA sequencing

Sequencing is the process by which the order of the bases along a single strand of DNA can be determined. Most techniques use a DNA replication reaction, though very recent technologies assess differences in size, shape, or charge of bases to determine the sequence. The sequences of any encoded proteins can be deduced based on the genetic code.

➔ *see also 9.4 Molecular biology techniques: genome sequencing (p. 239).*

1. Dideoxy chain termination sequencing (also known as Sanger sequencing)

- DNA to be sequenced is purified—if double-stranded, it is denatured prior to sequencing

- This DNA is used as template in a DNA replication reaction using dNTPs spiked with a low concentration of dideoxyribonucleotide triphosphates (ddNTPs) as random chain terminators (lack 3′ OH on ribose)

- Reactions contain:

    i. purified DNA template

    ii. a synthetic single-stranded short (15–18 nt) oligonucleotide **primer** complementary to a known region of the DNA (e.g. the vector adjacent to the cloning site)

    iii. mixes of all four deoxynucleotide triphosphates (i.e. dATP, dTTP, dCTP, and dGTP)

    iv. a low concentration of the dideoxy form of each nucleotide triphosphate (ddNTP)—one per reaction in radiolabelled systems, all four together in modern fluorescence-based sequencing

    v. a modified form of *DNA polymerase* that will incorporate ddNTPs as well as normal dNTPs

    vi. radiolabelled dATP in traditional radiolabelled sequencing reactions OR ddNTPs each labelled with a different coloured fluorophore in modern fluorescence reactions

- A new DNA strand complementary to the template DNA is synthesized by *DNA polymerase* elongating from the 3′ OH of the primer

- DNA synthesis stops when a dideoxy nucleotide has been incorporated, as there is no 3′ OH group for the nucleophilic attack required to attach a new nucleotide to the growing DNA chain, giving a range of different sized fragments
  - ➔ *see 4.1 DNA replication (p. 81) for phosphodiester bond formation*
- DNA fragments are separated by size on urea–polyacrylamide gels or in thin glass capillary tubes—also termed 'sequencing by separation' (see Figure 9.3A)
- DNA bands are detected on X-ray film (for radiolabelled reactions) or by laser excitation as bands pass a detector (fluorescence sequencing, e.g. ABI Prism)
- The sequence is read according to the position that DNA fragments migrate
- 400–1,500 nucleotides can be sequenced per reaction

2. Pyrosequencing (also known as 454 sequencing)

- DNA is sequenced in real-time by directly analysing incorporation of new dNTPs during DNA synthesis—hence also known as 'sequencing by synthesis'
- DNA template (with primer annealed) is immobilized in a nano flow cell with a light detector
- Reaction mix includes *DNA polymerase, ATP-sulfurylase,* and *luciferase,* buffer, and Mg^{2+}
- Each dNTP in turn is flowed over immobilized template, with washing in between
- When a nucleotide is incorporated into the DNA:
  - i. pyrophosphate is released
  - ii. PP_i is converted to ATP by *ATP-sulfurylase*
  - iii. *luciferase* uses the ATP to generate light
  - iv. amount of light given off is proportional to the number nucleotides incorporated when one type of dNTP is flowed through the cell (e.g. incorporation of GG will give off twice as much light as a when a single G is incorporated) (see Figure 9.3B)
- The next dNTP is then passed through the flow cell, and so on. This is an extremely rapid reaction using tiny volumes of reagents (nanolitres)
- An alternative uses fluorescent dNTPs—incorporation is detected using a charge-coupled device (CCD) after laser excitation; the fluorophore is then cleaved prior to addition of the next dNTP

3. Other DNA sequencing methods

- An older sequencing method relies on **chemical cleavage** of the phosphodiester bond using reagents specific for cleavage at purines or pyrimidines (Maxam and Gilbert). It is not widely used
- Other sequencing techniques in development include use of electron microscopy and nanopores
- The high speed and low cost of modern solid state sequencing should soon allow regular sequencing of individual genomes, e.g. to determine correct treatment of individual patients—pharmacogenomics ('the $1000 genome')

DNA manipulation and analysis

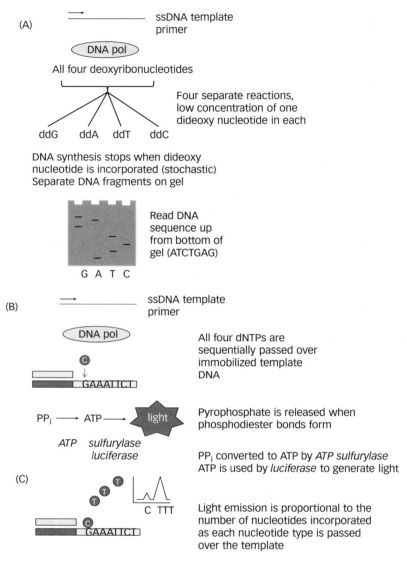

Figure 9.3 DNA sequencing. (A) The dideoxy reaction with gel separation of fragments. (B, C) Pyrosequencing.

DNase1 footprinting analysis

This is a technique to identify the region or sequence of DNA to which a protein binds.

- Two parallel reactions are set up, both with identical DNA which is radiolabelled on one 5′ end
- The protein of interest is added to one DNA sample and allowed to bind (it can be a purified protein or a mix of proteins such as a total cell lysate)

- *DNase1* nuclease is added: it will cleave DNA at regular intervals unless the DNA is 'hidden' by the protein bound to it; protein binding can distort the DNA helix and create *DNase1* hypersensitive sites as well as masking some sites
- DNA fragments are separated on acrylamide gels (as in DNA sequencing) to separate DNA fragments very similar in size
- Where the protein is bound to the DNA, DNase1 cannot cleave and so the protein leaves a 'footprint'—this is seen as a region of the gel with an absence of DNA fragments (see Figure 9.4)
- DNA sequencing reactions can be run in parallel on the same DNA region to identify which sequences are bound by the protein
- *In vivo* footprinting can also be used to find where endogenous proteins bind to DNA within cells
- Variations can use protection from chemical modification, e.g. copper phenanthroline modification/protection studies

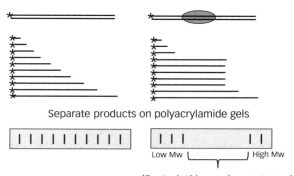

Separate products on polyacrylamide gels

Low Mw   High Mw

'Footprint' i.e. region protected from
DNase 1 cleavage by bound protein

Figure 9.4 DNase1 footprinting. Two identical DNA samples (radioactively labelled on the 5′ end of one strand, shown by asterisk) are incubated in the presence or absence of protein, then exposed to *DNase1*. The enzyme is blocked from cleaving DNA where protein is bound. Products of cleavage are analysed on polyacrylamide gels; absence of bands represents the 'footprint' of the protein, i.e. the position where it bound to the DNA.

Polymerase chain reaction (PCR)

This reaction allows the region of DNA between two known sequences to be specifically amplified in an exponential manner for subsequent identification, cloning, or sequencing. Very little DNA is required as starting material.

Contents of a PCR reaction

- Template DNA (the target may be a shorter region within the template, e.g. a gene within a genome)

DNA manipulation and analysis

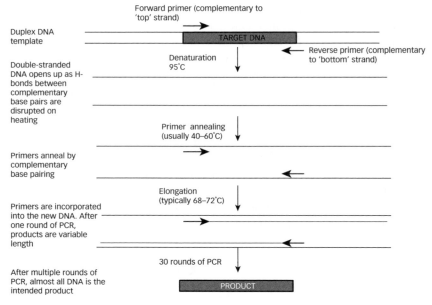

Figure 9.5 Standard PCR. Primers are depicted in bold for emphasis.

- Two single-stranded synthetic oligonucleotide primers complementary to sequences on the template either side of the target (Figure 9.5):
  - i. primers are usually 18–30 nucleotides each in length
  - ii. one primer must be complementary to the 'top' strand and the other to the 'bottom' strand, such that each strand can be copied starting from a bound primer
- All four dNTPs
- Thermostable *DNA polymerase* derived from thermophilic (heat-loving) bacteria, especially *Taq polymerase* from *Thermophilus aquaticus* (from hot springs)
- Buffer with pH 7–8 and $MgCl_2$ (0.5–3 mM)
- Other components may be added to deal with template secondary structures, e.g. formamide
- NB Primers, dNTPs, and *DNA polymerase* (plus Mg ions and buffer) are present in vast molar excess over template; primers and dNTPs are incorporated into the new DNA

Steps in PCR

- Template DNA is denatured (made single-stranded) at high temperature, typically ~95°C
- Primers are annealed to the template—temperature used is related to T_m of primers (range usually 40–60°C)[1]

---

[1] T_m is the temperature at which 50% of the DNA will be melted, i.e. single stranded. For short primers, it can be calculated roughly as $T_m = 4(G+C) + 2(A+T)$. The annealing temperature used in PCR is approx. $T_m - 5°C$.

- New DNA is synthesized complementary to the sequence of the template DNA at 68–75°C (according to optimum of the *DNA polymerase* used)
- Denaturation, annealing, and polymerization steps are repeated many times—usually about 30–40 cycles (one molecule of DNA is amplified to over a billion copies in 30 cycles)

Features of PCR

- Any DNA can be used as template, e.g. plasmids, genomic or cDNA libraries, total genomic DNA derived from individual cells, etc.
- Primers are designed according to known DNA sequence either side of fragment to be amplified, or according to regions of greatest homology between genes from different species, to allow cloning of the gene from a new species
- Primers can also be designed from protein sequence information using the genetic code:
    - i. in this case, primer sets must include all potential codons for a single amino acid—**degenerate primers** (alternatively, inosine can be used to base pair with any of the four dNTPs)
    - ii. degree of primer degeneracy will determine the specificity of the PCR reaction: as long as base pairing with the template is sufficiently selective at the annealing temperature, homologous or 'similar' DNAs will be isolated. More degenerate primers have a greater chance of annealing to less related sequences elsewhere in the template DNA
- PCR reaction amplifies specific sequences between two primers **exponentially** by repeating the DNA polymerization procedure (non-specific DNA may be amplified linearly)
- DNA fragments generated by PCR can be inserted into a vector or sequenced directly (cycle sequencing)
- New restriction enzyme sites, or other useful sequence motifs (e.g. epitope tags) can be added by using primers containing the required DNA sequences
- Since PCR can amplify large amounts of a specific DNA region from very small amounts of starting DNA, it is useful in forensic science and embryo testing
- **Inverse PCR** allows amplification of unknown DNA both sides of a known sequence by cyclising the DNA (see Figure 9.6)

Uses of PCR

- Isolation of novel genes or cDNAs from mixed cDNA preparations, or even genomic DNA, often circumventing the need to screen libraries
- Generation of probes for library screening
    - ➜ *see 9.3 Methods for cloning genes of interest (p. 237)*
- Preparation of relatively large amounts of specific fragments of linear dsDNA
- Altering DNA sequences, e.g. site-directed mutagenesis, adding restriction enzyme sites, forming specific deletion constructs, etc.
- Genotyping—using allele-specific primers

DNA manipulation and analysis

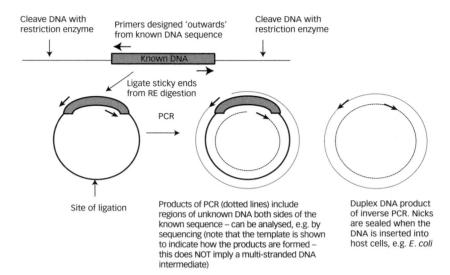

Figure 9.6 Inverse PCR.

- Analysing presence of DNA bound to protein, e.g. in chromatin immunoprecipitation (ChIP)

  ➔ *see 9.11 Immunological techniques (p. 257) for ChIP*

Looking for extra marks?

Consider the uses of PCR in clinical diagnostics, paternity testing, and forensics (e.g. can be used to amplify tiny amounts of DNA found in blood, hair or semen at crime scenes prior to DNA fingerprinting).

Site-directed mutagenesis

Site-directed mutagenesis is used to alter individual bases of DNA.

- Can be used to alter the coding sequence of DNA, hence changing the amino acid sequence of the encoded protein, e.g. in engineering enzyme active sites to bind substrate more tightly, or mutation of potential phosphorylation sites (serine, threonine, or tyrosine) to non-phosphorylatable residue (e.g. alanine) to test effect of phosphorylation on protein function
- Steps in site-directed mutagenesis (see Figure 9.7):
  i. gene/cDNA to be altered is cloned into a circular vector
  ii. synthetic oligonucleotide primers are designed carrying one or several mismatched bases to span the DNA sequence to be altered, in opposite orientations (like inverse PCR—see Figure 9.6)—the altered nucleotide(s) must be near the centre of both primers to ensure efficient annealing

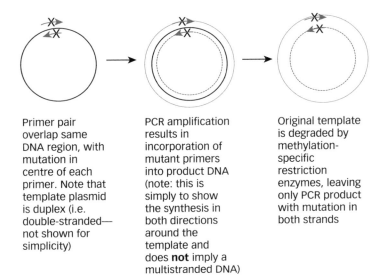

| Primer pair overlap same DNA region, with mutation in centre of each primer. Note that template plasmid is duplex (i.e. double-stranded— not shown for simplicity) | PCR amplification results in incorporation of mutant primers into product DNA (note: this is simply to show the synthesis in both directions around the template and does **not** imply a multistranded DNA) | Original template is degraded by methylation-specific restriction enzymes, leaving only PCR product with mutation in both strands |

Figure 9.7 Site-directed mutagenesis by PCR.

   iii. PCR amplification from the primers synthesizes the full circular DNA of the vector – each new duplex contains the altered nucleotide as the primer is integrated into the product
- Since the PCR products are amplified exponentially, the vast majority of product contains the altered base
- The PCR method is far quicker and easier than earlier methods of site-directed mutagenesis

9.2 cDNA, VECTORS, AND LIBRARIES

Complementary DNA (cDNA)

Complementary DNA is a DNA copy of an RNA molecule. cDNA constitutes the coding sequence of a gene but does not contain promoter sequences or introns. cDNA is sometimes used to describe a DNA strand complementary to a DNA strand of interest.
- mRNA cannot be cloned—must first be converted to DNA
- One strand of DNA complementary to the mRNA is generated by an RNA-dependent DNA polymerase *reverse transcriptase (RT)* (see Figure 9.8)
- *RT* requires:
   i. an RNA template
   ii. a primer—usually synthetic oligo-dT, which is complementary to the 3′ poly(A) sequence of eukaryotic mRNAs, or random hexamer oligonucleotides
   iii. dNTPs + buffer

cDNA, vectors, and libraries

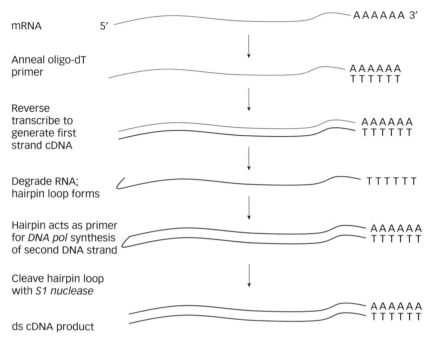

Figure 9.8 cDNA synthesis.

- Once a single strand of DNA has been synthesized, the mRNA is then degraded by chemical or enzymatic means, leaving 'first strand cDNA' with a hairpin loop at one end
- The single strand of DNA can be converted to double-stranded cDNA by *DNA polymerase*, using the hairpin loop as a primer for second strand synthesis
- The hairpin is cleaved by treatment with *S1 nuclease*
- The fidelity of *reverse transcriptase* is lower than *DNA polymerases* as it lacks proofreading activity
- Processivity of *reverse transcriptase* (*RT*) is low—hence cDNAs are often shorter than the mRNA template and lack 5′ sequences as *RT* starts from the 3′ end of mRNA template
- Use of cDNA instead of genomic DNA overcomes problems with expressing eukaryotic genes in bacteria, as cDNA lacks introns and eukaryotic regulatory sequences which bacteria cannot process
- Can be used to generate cell-type specific collections of cDNAs by using the whole population of mRNAs purified from a particular cell type. Because of different expression patterns, a collection of cDNAs from one cell type may differ from a cDNA collection from a different cell type

Vectors

Vectors are made of DNA and are used to 'carry' DNA directionally, e.g. into a specific host cell. There are different types used in molecular biology, according to both the ease of manipulation and amount of foreign DNA that can be accommodated in the vector. Vectors usually have restriction sites suitable for insertion of foreign DNA, a selectable marker, and a replication origin to ensure their propagation in the host cell. See Table 9.1 for summary of vectors commonly used in molecular biology.

1. Plasmid vectors

- Usually small circular bacterial DNA molecules (~2–15 kb) that can carry foreign DNA fragments for amplification or expression
- Contain an **origin of replication** and so will replicate in *E. coli* without integrating into the genome—essential to ensure high **copy number** (large number of plasmids per cell) and hence amplification of the inserted DNA sequence
- Contain **selectable marker**(s) so that *E. coli* transformed with a plasmid will survive in circumstances that kill non-transformed cells. Common selectable markers include:

    i. antibiotic resistance genes, e.g. tetracycline, ampicillin

    ii. fluorescence markers can be used together with antibiotic resistance, e.g. in generation of transgenic animals

- Plasmids engineered for use in molecular biology applications contain many restriction enzyme cleavage sites clustered in one region (**multicloning site or MCS**). MCSs allow cloning of many different foreign DNA fragments created by cleavage with a range of restriction enzymes—a variety of sticky ends can be created in the vector, compatible with the overhangs of the fragments to be cloned
- **Expression vectors** also contain a promoter and **Shine–Dalgarno sequence** (ribosome binding site), plus transcriptional terminators and translation stop codons, to allow correct transcription and translation of the foreign DNA
- For key features of a plasmid vector see Figure 9.9

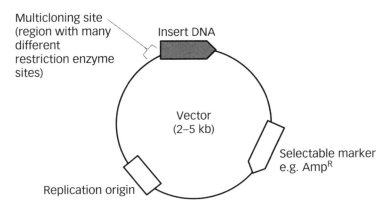

Figure 9.9 Features of a typical plasmid vector.

2. Variations on plasmid vectors

- Designed so that inserted DNA disrupts a functional bacterial gene to facilitate selection:
    i. e.g. tetracycline resistance—*E. coli* transformed with these plasmids are resistant to ampicillin (encoded on a different region of the plasmid) but killed by tetracycline
    ii. e.g. *lac Z* (encodes *β-galactosidase*)—*E. coli* with functional *β-galactosidase* convert X-gal substrate from colourless to blue; those with foreign DNA inserted into the lac Z gene appear white, allowing 'blue/white' selection
- Vectors can contain extra selectable markers, and both bacterial and eukaryotic origins of replication, to allow the vector to be 'shuttled' between *E. coli* and another organism—these are called **shuttle vectors**:
    i. e.g. neomycin resistance in mammalian cells
    ii. e.g. amino acid biosynthetic enzyme (his2, ura3, etc.) in yeast

3. Other vectors

- **Bacteriophages** are viruses that naturally infect bacteria and so serve as very efficient delivery vehicles for getting foreign DNA into bacteria—16-fold higher cloning efficiency than plasmid vectors, e.g.
    i. phage M13—circular genome, has single-stranded and double-stranded DNA phases of life cycle; takes only small inserts of foreign DNA
    ii. phage λ—double-stranded linear DNA genome, accepts larger DNA inserts (insertion vectors take a few kb and stuffer vectors take up to ~30 kb)
- **Cosmids** and **fosmids** can take large amounts of insert DNA (~ 40 kb) but are based on plasmids—widely used in genome projects (e.g. *Caenorhabditis elegans* genome library is cloned in fosmids)
- **Bacterial artificial chromosomes** (BACs)
    i. possess bacterial origins of replication
    ii. selectable markers
    iii. can contain very large DNA inserts (100s of kb)
- **Yeast artificial chromosomes** (YACs) are linear and have centromeres, telomeres, and replication origins as well as an auxotrophic selectable marker—can accept Mb of DNA
- Viruses that infect insect cells, e.g. **baculovirus**—used for protein expression
- Modified forms of viruses that infect mammalian cells e.g. **retroviruses** (HIV) and **adenovirus** (AdV) are useful for expressing genes in human cells and hence are used in **gene therapy**

Libraries

A library is a comprehensive collection of DNA fragments that have been inserted into vectors.

Vector	Host	Vector structure	Insert size (kb)
Plasmid	*E. coli*	circular plasmid	1–5
M13	*E. coli*	circular virus	1–4
Lambda	*E. coli*	linear virus	2–25
Cosmid	*E. coli*	circular plasmid	35–45
Fosmid	*E. coli*	circular plasmid	35–45
BACS	*E. coli*	circular plasmid	50–300
YACS	*S. cerevisiae*	linear chromosome	100–2000

Table 9.1 Vectors commonly used in molecular biology Phage λ DNA can cirularize on entry into host bacterial cell via 12 nt cohesive ends (cos sites).

1. Genomic libraries

- Made up of restriction enzyme-generated fragments of the genomic DNA of an organism (e.g. human), inserted into a vector cut with same restriction enzyme
- DNA fragments will include promoter sequences, introns, and other non-coding DNA, in addition to coding sequences
- Genomic libraries are specific for a particular organism but do not usually differ between different cell types of that organism (same genomic DNA in human liver as in human brain)

2. cDNA libraries

- Comprise vectors containing cDNA copies of virtually all the mRNA molecules present in an individual cell type (e.g. liver hepatocyte)
- Will vary between cell types, depending on the subset of genes transcribed into mRNA in each cell type, since the cDNA is generated from mRNA (e.g. human liver cDNA library is very different from a human brain cDNA library)
- Contain only the coding regions, and 5′ and 3′ untranslated sequences, not promoters or introns or intergenic sequences
- Many cDNAs in a library will contain an intact 3′ end of the coding sequence but will not necessarily be full length (i.e. possess the 5′ end) as *reverse transcriptase* has limited processivity and tends to 'drop off' the RNA template before reaching the 5′ end

3. Expression libraries

- Designed for expression of foreign DNA in, for example, bacterial host cells—hence only cDNA because bacteria cannot process transcripts from eukaryotic genomic DNA due to the presence of introns, regulatory sequences and non-coding DNA
- cDNAs are cloned downstream from an inducible promoter (e.g. from the lac operon) and ribosome binding site, in an expression vector
- *E. coli* bacteria are transformed with the library
- Expression of the foreign DNA is induced by addition of a specific inducer, e.g. IPTG for vectors under the control of the lac operon

  ➔ *see 4.2 RNA synthesis (transcription) (p. 94) for transcriptional regulation of the lac operon*

- The bacteria transcribe and translate the foreign DNA, producing the encoded protein product which can be either:
  - i. soluble
  - ii. present in **inclusion bodies**
  - iii. secreted (usually requires a secretory signal fused in frame with the cDNA)

9.3 METHODS FOR CLONING GENES

Now that many genome sequences are known, most new genes are cloned by PCR directly from cDNA or genomic DNA (gDNA). Where the genome sequence is unavailable, PCR can still be used with degenerate primers, or traditional library screening is used. Prior to PCR and genome sequence information, many genes were isolated by their ability to correct a specific phenotype, or by their co-inheritance with genetic markers.

Cloning by PCR

- Primers are designed according to genome sequence. If not known, regions of maximal homology are determined by sequence comparison of the protein or DNA in species from which the gene has already been cloned—using bioinformatics software, e.g. BLAST
- Oligonucleotide primers are synthesized:
  - i. these are used directly as primers in a PCR reaction with template DNA (fragmented gDNA, newly synthesized single-strand cDNA, or a gDNA or cDNA library)
  - ii. if nucleotide sequences vary widely between different species, inosine can be used in the synthetic oligonucleotide as it will hybridize with all four bases of DNA
  - iii. degenerate primers can be used if the genome sequence is unknown but protein sequence is known (e.g. using Edman degradation)
    - ➔ *see 9.10 Protein analysis (p. 252)*
  - iv. restriction enzyme sites can be included to facilitate subsequent cloning steps
- If the cDNA or gene is too long to be amplified in a single PCR reaction, several adjacent regions can be amplified in individual PCR reactions (preferably using primers spanning naturally occurring restriction enzyme sites):
  - i. PCR products are cleaved with the restriction enzyme to produce compatible ends (usually easier after cloning into an entry vector)
  - ii. restriction fragments are ligated together to generate the full length construct.
- PCR products are cloned into 'entry' vectors, either with T overhangs (if PCR used *Taq*) or blunt ends (for proof-reading polymerases) and can then be sequenced

Cloning a gene according to protein sequence

- The protein is purified according to activity, or as a band on a sodium dodecyl sulphate polyacrylamide gel (SDS-PAGE)
- Partial amino acid sequence is determined:
    i. Edman degradation is generally used
    ii. more sensitive techniques such as mass spectrometry allow amino acid determination from much smaller amounts of protein
- The sequence of the DNA encoding these amino acids is deduced, although the degeneracy of the genetic code means that there are always several possible DNA sequences.
- This DNA sequence is synthesized *in vitro* as a set of degenerate oligonucleotides that includes all possible variations of the genetic code for each amino acid
- The oligonucleotide is radioactively labelled and used as a probe in library screening
- Alternatively, the oligonucleotide can be used as a primer PCR to for direct cloning, or to generate a larger, more specific probe for library screening

Cloning by library screening

- *E. coli* are transformed with cDNA or genomic library, and grown on selective media so only bacteria that have taken up the vector grow
- Bacterial colonies are sampled by placing a nitrocellulose filter on top of the colonies, noting its orientation on the plate, then lifting it off; cells are lysed to release DNA
- DNA that has stuck to the filter is fixed by UV cross-linking or baking
- The filter is incubated with a probe, i.e. specific DNA or RNA sequence of interest that is labelled (radioactive, fluorescent or otherwise tagged)—this is a probe:
    i. the longer the probe, the lower the chance of random cross-hybridization
    ii. different salt, temperature, and detergent conditions can be used to wash the filter once the probe has hybridized—the higher the temperature or the lower the salt concentration, the more **stringent** the wash
    iii. at high stringency, only sequences very similar to the probe will be bound; at lower stringency, less closely related sequences can bind—useful for isolating genes from evolutionarily distant species
- The positions where the probe hybridizes with DNA fixed on the filter (identified, for example, by exposure to X-ray films) are used to identify bacterial colonies containing the DNA of interest ('positives')
- Positive *E. coli* colonies are picked and grown up in bulk to amplify plasmids containing the DNA insert of choice
- Vector DNA is purified and sequenced in the region of the foreign DNA insert
- Several rounds of screening are usually required to eliminate false positives
- See Figure 9.10 for library screening strategy
- VARIATION: if an antibody specific to the protein is available, this can be used instead of a DNA probe to screen an expression library

Methods for cloning genes

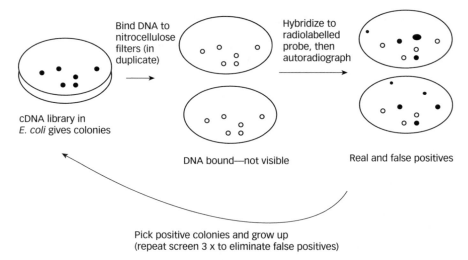

Figure 9.10 Library screening.

Cloning by phenotype

- A cell line defective in a specific gene function is used, e.g. human disease xeroderma pigmentosum
  - → *see Chapter 4: Nucleotide excision repair (p. 125)*
- These cells are transfected with a cDNA expression library in shuttle vectors
- Correction of defective phenotype is assessed—this is called a **complementation assay**
- Plasmid vector DNA (with insert) is isolated from transfected cells showing the corrected phenotype
- Isolated DNA is then amplified by transformation into *E. coli* and growth in large-scale culture, then purified and sequenced
- This process is more frequently used to isolate higher eukaryotic homologues of yeast genes by complementation of the relevant yeast mutant phenotype

Positional cloning

- A complex procedure involving analysis of large regions of chromosomes which are mapped by analysing loci that are present/absent in families with specific diseases, e.g. cystic fibrosis—**physical mapping**
- Genetic markers are identified which co-segregate with the disease phenotype
- Once a locus which correlates with the disease has been identified, overlapping contiguous genomic fragments (contigs) are isolated to 'walk' along the chromosome until the gene is reached
- The conservation of sequences in that region is determined between related species (e.g. human and mouse), since this implies the region is important as it has been retained through evolution

- Verify that conserved regions encode appropriate mRNAs (e.g. gene for cystic fibrosis must be expressed in the lung)
- Sequence the genes that are appropriately expressed and compare sequence between patient and unaffected family members

9.4 GENOME SEQUENCING

The genomes of many organisms, including the human genome, have now been sequenced. Strategies have been developed to decipher genome sequences by breaking the genome down into small 'sequenceable' chunks, complemented by new bioinformatics techniques designed to allow alignment and assembly of the sequence information into a complete genome sequence, and to deal with the enormous amount of sequence information generated.

Iterative sequencing (e.g. Human Genome Consortium project)

- Chromosomes are fragmented and fragments are inserted into BAC or YAC libraries (inserts up to 1,500 kb).
- Such fragments will by chance overlap and represent the whole of the chromosome as overlapping contiguous segments (contigs)
- For each BAC or YAC, a series of fragments of approximately 40 kb is generated by restriction enzyme digestion and inserted into cosmids
- The insert of each cosmid is fragmented into 1–10 kb fragments and inserted into M13 or plasmid cloning vectors
- The insert of each plasmid/M13 vector is short enough to be sequenced directly by dideoxy chain termination using primers complementary to the vector DNA (see Figure 9.3)
- The sequence information is built up by comparing sequence and finding overlaps. Cosmids with the minimum overlap are chosen and sequenced in both directions (again through fragmenting, inserting fragments into plasmids and directly sequencing)—this is known as 'cosmid walking'
- Bioinformatics is used to piece together overlapping sequence information and assemble the contigs to form a complete genome sequence
- See Figure 9.11 for steps in interative genome sequencing

Shotgun sequencing

- Chromosomes are fragmented and inserts cloned into BACs
- Both ends of each BAC are sequenced directly to give a 'BAC fingerprint'
- A 'seed BAC' is chosen and the 150 kb insert is fragmented and inserted into plasmid or M13 libraries

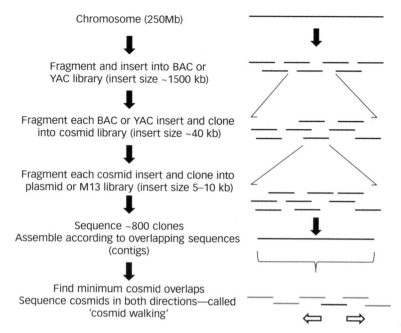

Figure 9.11 Genome sequencing.

- Each insert is sequenced and sequence information is assembled using bioinformatics
- These are compared with BAC fingerprints
- Contiguous BACs with minimal overlap are sequenced (via fragmentation and insertion into plasmid or M13 libraries)—this is known as 'BAC walking'

New generation genome sequencing (NGS)

- Multiple parallel arrays of solid-state pyrosequencing (e.g. 454) or fluorescence sequencing
- Allows reads of 400—600 Mb in a 10-hour run on one machine
- Uses very low volumes per reaction (whole genome in <10 μl)
- Other technologies measure tiny changes in electrical charge as DNA passes through a nanopore—very rapid and low cost
- Has significantly reduced the time taken and cost—a '$1000 genome' is predicted to be a reality soon

9.5 GETTING DNA INTO CELLS

Once the DNA of interest has been cloned, it is usually necessary to insert it into living cells (e.g. *E. coli*) for plasmid amplification, recombinant protein expression, or into eukaryotic cells for studying gene activity or protein function.

Transforming *E. coli*

- Uptake of foreign DNA into *E. coli* is called **transformation**
- *E. coli* can be made competent to take up foreign DNA by chemical treatment (see Figure 9.12):
  - i. bacteria are incubated on ice with DNA in the presence of RbCl and $CaCl_2$—these help to neutralize the negative charge on the DNA and the cell membrane, and the low temperature reduces membrane fluidity
  - ii. DNA associates at the membrane in regions known as adhesion zones
  - iii. cells are exposed to transient heat shock (at ~42°C for <1 minute)—this creates a thermal gradient across the membrane that encourages DNA uptake
  - iv. cells must be allowed to recover (e.g. incubate at 37°C for 30 minutes) before antibiotic selection is imposed
- Alternatively, electroporation can be used whereby a very high voltage is applied across the membrane for very short time periods, encouraging DNA uptake
- Suitable *E. coli* strains must be used, e.g. lacking *nucleases* (for DNA amplification) or *proteases* (for recombinant protein expression)
- Some host strains have phage *T7 RNA polymerase* encoded on the bacterial genome under control of the lac operon to allow tight regulation of expression of foreign cDNAs from the T7 promoter present on the vector

Getting DNA into mammalian cells

- Uptake of foreign DNA into mammalian cells is called **transfection**
- Mammalian cells take up liquids from their medium by micropinocytosis—they can also take up tiny particles through this route
  - i. DNA is incubated with calcium phosphate (which precipitates) and this precipitate is then taken up into cells (with low efficiency)
  - ii. efficiency can be improved by using liposomes or polyamines
- Electroporation can be used to stimulate uptake of foreign DNA into mammalian cells

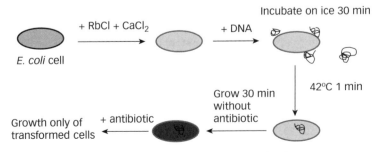

Figure 9.12 Bacterial transformation by the heat shock method.

- Viruses are often used to carry DNA into mammalian cells—retroviruses (e.g. derivatives of HIV) or DNA viruses based on adenovirus
- DNA can be microinjected directly into mammalian cells—this is used especially with large cells such as eggs, e.g. in making transgenic animals

Biolistics

- This technique directly introduces DNA into cells and even organelles using a particle gun—also known as the 'gene gun' (see Figure 9.13)
- Tiny metal particles (microprojectile) are coated with DNA and suspended in a drop on a macroprojectile
- An explosive discharge forces the macroprojectile forwards, where it is stopped suddenly by collision with a stopping plate
- The metal particles coated with DNA continue forwards and are forcefully injected into the tissue held in front of the stopping plate
- Has been used successfully to:
  i.   get DNA into plant cells (cell wall prevents other transfection methods and removal of the cell wall makes the plant cell osmotically and mechanically very fragile)
  ii. create transgenic worms
  iii. get genes or cDNA into plastids and mitochondria
  iv. a variant is used for needle-free vaccination

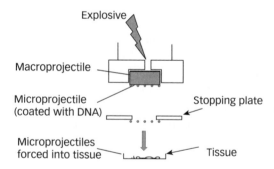

Figure 9.13 Biolistics 'gene gun' in cross-section. Gold beads are coated with DNA (forming the microprojectile), then placed on the macroprojectile. A controlled explosion pushes the macroprojectile into a stopping plate; the microprojectiles continue at high velocity and penetrate the tissue in the dish below.

9.6 GENE THERAPY

The aim of gene therapy is to replace a defective disease gene with a fully functioning wild-type version.

- Suitable for treatment of diseases caused by mutation of a <u>single</u> gene, e.g. Duchenne muscular dystrophy, cystic fibrosis (CF), severe combined immunodeficiency (SCID), and Parkinson disease
- The wild-type gene is cloned and delivered either on a plasmid or within a virus
- Gene delivery:
    i. retroviruses provide effective gene delivery as they integrate into the host genome, but they can cause cancer if they insert next to an oncogene
    ii. DNA viruses such as adenovirus can be used but they can trigger a severe immune response
    iii. DNA can be delivered directly in liposomes but efficiency is low
- Some cases exist where diseases have been cured (e.g. SCID), some positive benefits (CF), but progress is slow
- In some cases of polygenic disease where one factor is particularly important, it is possible to restore that factor by gene therapy, e.g. Advexin gene therapy restores the tumour suppressor p53 in cancers lacking p53

9.7 GM CROPS

The aim of genetically modifying crops is to increase yield, or to introduce a new nutrient or other useful trait such as pest resistance into a plant.

- Plant-specific vectors can be used, e.g. those derived from *Agrobacterium tumefaciens* (for dicotyledons)
- Alternatively, DNA can be 'shot' into plant cells on tiny gold particles using biolistics
- Examples include:
    i. golden rice (increased production of vitamin A to prevent childhood blindness)
    ii. Bt-cotton expressing *Bacillus thuringiensis* toxin that prevents crop destruction by insect pests, and reduces the need to use harmful chemical pesticides, so improving health of farmers (introduced in 1996)
    iii. attempts to increase the efficiency of photosynthesis to enhance overall crop yields

9.8 RNA MANIPULATION AND ANALYSIS

RNA isolation

- Total RNA can be purified on chemical resin and any residual DNA removed by treatment with *DNase*
- Messenger RNA can be isolated from eukaryotic cells by virtue of the polyA tail at its 3′ end.

i. polyA$^+$ mRNA binds to synthetic oligo-dT oligonucleotides that can be immobilized on columns—other RNAs lacking a polyA tail (e.g. rRNA, tRNA) do not bind and can be washed off the column

ii. mRNA is eluted using an excess of oligo-dT or by increasing salt concentrations

RNA concentration determination

- Absorption of UV light at wavelength 260 nm is measured
  i. RNA concentration is given by: $40 \times A_{260}$ = RNA concentration in μg/ml
  ii. e.g. an optical density of 1.0 at 260 nm indicates an RNA concentration of 40 μg/ml
- Purity is assessed by measuring absorption also at 230 and 280 nm (peak absorbance wavelengths for carbohydrates and proteins, respectively). A ratio of 1:2:1 at A_{230}:A_{260}:A_{280} indicates the RNA is pure

Northern blotting

A method for determining the expression levels and transcript size of an RNA product of gene of which at least some of the sequence is known.
- Total RNA is extracted from cells and separated on an agarose gel in the presence of formamide
- RNA is transferred from the gel to nitrocellulose/nylon membrane by capillary action in neutral buffer (similar to Southern blot)
- The membrane is baked or UV cross-linked to fix the RNA onto the membrane
- Single-stranded DNA probe complementary to the RNA of interest is incubated with the membrane—probe is radiolabelled or fluorescently tagged
- Unbound probe is washed off and the blot is exposed to X-ray film—a band is observed where the probe has bound to the blot (see Figure 9.14)
- Northern blots can be used quantitatively to measure the level of gene expression
- The size of the transcript can be determined by the position the band has migrated on the gel compared with markers—this allows determination of transcript size and of any alternative splice variants

RT-PCR

Reverse transcription PCR is used to amplify large amounts of DNA corresponding to an individual RNA species.
- The RNA template (total RNA or purified mRNA) is incubated with *reverse transcriptase* (*RT*), primers (random hexamers or oligo-dT), dNTPs, Mg^{2+}, and buffer
- First strand cDNA is synthesized: *RT* generates a short hairpin at one end

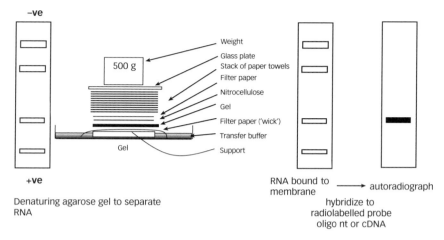

Figure 9.14 Northern blotting.

- Primers, dNTPs, etc. are removed in a 'clean-up' step and the first strand cDNA is used as template for subsequent PCR using gene-specific primers

  ➔ *see 9.1 DNA manipulation and analysis (p. 227) for PCR*
- Alternatively, gene-specific primers can be employed complementary to the transcript sequence:
    i. these act both in the RT step and in PCR, and so there is no need to clean up prior to starting the PCR reaction
    ii. Both *reverse transcriptase* and *Taq* can be present from the outset as the RT reaction optimal temperature (e.g. 50°C) is far below that for *Taq* (72°C); incubation at 50°C for 30 min allows RT to act
    iii. Incubation at 95°C during the first PCR step denatures *reverse transcriptase*, so only PCR amplification of DNA occurs
- Quantitative RT-PCR can be used to measure the amount of starting RNA—uses fluorescent intercalator, e.g. SYBR green

Microarrays

These provide a powerful method for comparing relative levels of expression of >10,000 genes at a time on a single glass slide.

- cDNAs or oligonucleotides are printed onto a glass slide (a chip) using a modified inkjet printer that uses DNA rather than ink (many microarrays are commercially available)
- To test relative gene expression, cDNA is prepared from mRNA extracted from different types of cells, or the same cells exposed to different conditions, e.g. ± drug, cancer vs. normal, old vs. young, etc.
- Test (T) cell cDNA will be compared with reference (R) cell cDNA

RNA manipulation and analysis

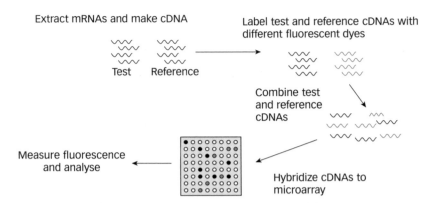

Figure 9.15 Microarray analysis for comparative gene expression studies.

- T and R cDNAs are labelled with different fluorescent tags (e.g. red for test, green for reference)
- T and R cDNAs are hybridized to chip: if gene is expressed, cDNA binds and fluoresces (see Figure 9.15):
  - i. T cDNA binding → red signal (gene expressed only on treatment)
  - ii. R cDNA binding → green signal (gene expressed only in control)
  - iii. T and R both bind → yellow (gene expressed under both conditions)
- Can measure both up- and downregulation of gene expression in cell of interest vs. reference cell

RNA interference (RNAi)

This method allows selective knockdown of gene expression in many eukaryotic organisms using short interfering RNA molecules. Regulation of gene expression by RNAi now appears to provide a significant level of endogenous gene expression control, as well as being a potential cellular mechanism to fight infection by double-stranded RNA viruses. RNAi is a powerful tool for probing eukaryotic gene function.

- Discovered in *C. elegans* using antisense RNA to decrease gene expression: the antisense gave some disruption, the control 'sense' RNA gave none, but surprisingly, even better gene downregulation was observed if both sense and antisense oligonucleotides used together (also in petunia colour regulation)
- Long double-stranded (ds) RNAs (>200 nt) found to silence gene expression in lower eukaryotes and plants
- Utilizes cellular RNA interference pathway (Figure 9.16):
  - i. dsRNAs cut into 20–25 nt siRNAs by *Dicer* (an *RNase III*-like enzyme)
  - ii. siRNAs assemble into RNA-induced silencing complexes (RISCs) containing *endoribonucleases*

   iii. siRNA strands unwind to form activated RISCs

   iv. RISCs bind complementary mRNA via homology to the sequence of the siRNA

   v. RISCs cleave cognate mRNA near the middle of the region bound by the siRNA strand, destroying the mRNA

- RNAi in mammals:

   i. dsRNA as short as 25 nt induces an 'interferon response'— defence against RNA viruses (e.g. flu) leading to non-specific inhibition of protein synthesis and cell death

   ii. necessary to bypass this response using shorter (~22 nt) (usually synthetic) oligoribonucleotides, siRNA

- Different types of RNAi:

   i. short interfering RNA (siRNAi) uses ds RNA oligonucleotides—transfected into mammalian cells (e.g. using lipofection)—effect is transient

   ii. short hairpin RNA (shRNAi) is expressed in cells (from *RNA pol III* promoter) as a hairpin duplex RNA that is then processed by endogenous *Dicer* to short ds RNA—stable expression, so long-term effects and does not induce the interferon response

   iii. micro RNA (miRNA)—expressed in cells (from RNA *pol II* promoter)— stable expression, so long-term effects

- NB miRNA provides an important endogenous cellular mechanism to regulate gene expression

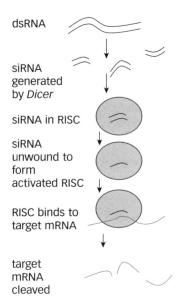

dsRNA

siRNA generated by *Dicer*

siRNA in RISC

siRNA unwound to form activated RISC

RISC binds to target mRNA

target mRNA cleaved

Figure 9.16 RNA interference in cells.

Check your understanding

What is the difference between a genomic library and a cDNA library? (*Hint: think about introns, regulatory sequences, etc. Be able to describe how each is made and what purposes they would be used for.*)

How does RNAi result in knock down of expression of specific target genes? (*Hint: see section above for details; remember to discuss length of siRNA and complementary base pairing between siRNA and target mRNA as this is crucial for specificity.*)

9.9 PROTEIN PURIFICATION

The aim of protein purification methods is to separate the protein of interest from a complex mixture such as a total cell lysate. It is a necessary first step in determining the structure and/or function of a particular protein. Proteins are normally purified on columns where the stationary phase comprises inert hydrophilic polymers, such as dextran ('Sephadex'), agarose, or polyacrylamide, normally in bead form. The chemistry of these polymers is modified to attract particular proteins or classes of protein.

Affinity protein purification

A rapid and efficient method for obtaining highly pure recombinant protein.
- cDNA encoding the protein of interest is cloned into an expression vector in frame with an appropriate affinity tag, e.g. hexa-histidine, *glutathione-S-transferase* (*GST*), etc.
- The vector with cDNA is transformed into a suitable *E. coli* host strain and protein expression is induced; cells are harvested by centrifugation after an appropriate time
- Cells are lysed by treatment with *lysozyme* and detergent, or by sonication, in buffer containing protease inhibitors
- The lysate is cleared by centrifugation then loaded onto an appropriate column:
    - i. nickel or cobalt for hexa-his-tagged protein
    - ii. glutathione for *GST*-tagged protein
- Tagged protein binds tightly to the column while untagged proteins flow through
- The column is washed and then the tagged protein eluted using:
    - i. high concentrations of imidazole (for his-tagged proteins)
    - ii. excess glutathione for *GST*-tagged proteins
- Some vectors encode a *protease* cleavage site between the tag and the protein of interest, so the tag may be cleaved off after purification. The tag can then be removed by passing down the affinity column, where the cleaved tag will bind but the purified protein will flow through
- For steps in purifying an *N*-terminally GST-tagged protein see (Figure 9.17)

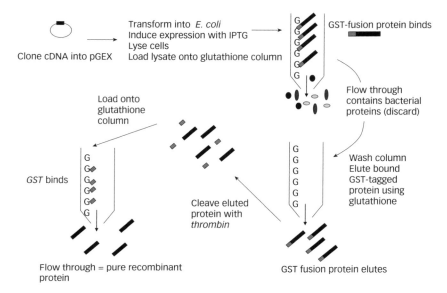

Figure 9.17 Affinity purification—example shown is of a GST-tagged protein in which the tag can be cleaved off by proteolysis.

- Can be combined with 'traditional' chromatography columns (e.g. size exclusion gel filtration) to increase purity
- Using immobilized antibodies against a specific protein as the stationary phase allows purification of a protein without a tag (immunoaffinity purification)
- Non-tagged proteins can sometimes be purified by their affinity for specific ligands:
    i. an enzyme may be purified by binding to its substrate on a column
    ii. glycoproteins can be purified on a stationary phase containing lectins
    iii. DNA binding proteins may be purified on heparin columns

Ion exchange chromatography (IEC)

A technique for separating proteins (and other biological molecules) according to their charge.

- The functional groups of proteins can have either positive and negative charge due to charged side chains (e.g. asp is negative, lys is positive); the sum of these provides a protein with an overall net charge
- A mixture of proteins (the sample) is loaded into a sample loop on a chromatograph
- A **mobile phase** (i.e. buffer) carries the sample onto a column that contains the **stationary phase** (a resin or gel of agarose or Sepharose beads to which are covalently bonded charged functional groups)

Protein purification

- Proteins with opposite charges to the column will bind:

  i. cation exchange chromatography retains cations (positively charged) since the stationary phase has negatively charged functional groups

  ii. anion exchange chromatography retains anions as the stationary phase has positively charged functional groups

- The binding of a protein to the resin can be altered by adjusting the pH and/or the ionic strength of the mobile phase— hence the bound protein can be subsequently eluted, e.g. by increasing the salt concentration

- Can be combined with other chromatography techniques (e.g. size exclusion gel filtration) to increase purity

Size exclusion (gel filtration) chromatography

A technique for separating proteins according to their size by passing down a column with pores of defined size. It can be used to distinguish between folded and unfolded proteins of the same species, and in determining quaternary structures of proteins (e.g. multimers).

- A column is packed with a stationary phase comprising very small porous polymer beads (e.g. acrylamide or Sepharose)—contains pores of different sizes, of similar dimensions to protein molecules

- The sample is loaded onto the column in buffer and runs through under a low flow rate

- Proteins equilibrate between the pores of the matrix and the space between the particles of the matrix—this equilibrium is determined by the mean span dimension which correlates closely with **hydrodynamic volume** (and **Stokes radius**) and so proteins are separated according to size:

  i. large proteins cannot enter the pores and elute rapidly from the column

  ii. smaller proteins enter the pores and elution is delayed

- Fractions are collected and analysed for protein content (e.g. by UV absorption)

- The pattern of elution of proteins is compared with a calibration curve of a mixture of proteins of known size that have been run through the same column to determine approximate molecular weight (plot log MW against V_e/V_o)

- The **exclusion limit** defines the upper limit of molecular weight that can be separated on the column—molecules above this size cannot be trapped in the stationary phase. The **void volume** (V_o) contains molecules too large to enter the pores

- The **permeation limit** defines the smallest size that can be distinguished on the column—molecules smaller than this enter the pores of the stationary phase completely and will elute as a single band

- Increasing column **length** increases its resolution

- Increasing column **diameter** increases its capacity

Thin layer chromatography

A quick and simple technique for separating components of a biological mixture.
- The stationary phase is finely divided silica or cellulose layered on a sheet of inert material
- Sample is applied in a small volume at one end of the plate
- Solvent moves up the plate by capillary action and the sample separates
- Useful in separating different phosphopeptides, lipids, nucleotides, etc.
- Used by Chargaff in determining ratio of nucleotides in DNA
  - ➔ *see Chapter 1: Key evidence for DNA structure (p. 21)*

Analytical gradient centrifugation

Proteins can be separated according to mass (and to some extent size and shape) by centrifugation though a gradient of a viscous substance such as sucrose or glycerol.
- A gradient is set up in a centrifuge tube with the highest concentration of sucrose/glycerol at the bottom and the lowest concentration at the top
- The sample is layered over the top of the gradient
- The tube is centrifuged at high speeds—the migration of a particle is a function both of its mass (the heavier the particle, the further it will move) and the resistance it encounters when moving through the liquid
- As particles move through the gradient, they encounter a liquid of increasing viscosity and density, and will migrate according to mass and how long they have been exposed to centrifugal forces—this can provide fine resolution between particles of different masses
- The gradient is fractionated and collected, and protein content analysed
- E.g. separation of ribosomal subunits

Reversed phase HPLC (high pressure liquid chromatography)

Used to separate peptides or other biologically relevant molecules under high pressure on densely packed columns.
- Stationary phase is a hydrophobic matrix with an aqueous (moderately polar) mobile phase:
  - i. stationary phase is very finely divided silica, normally modified with alkyl groups ('reverse phase')
  - ii. provides a high surface area/volume ratio, giving very high resolving power (industry standard) but requiring high pressures to drive the solvent
- Sample is injected onto the column and mobile phase is pumped through at high pressure
- Polar molecules elute rapidly, while more hydrophobic molecules are retained for longer on the hydrophobic column because of hydrophobic interactions

- Retention time is characteristic for a particular compound under standard conditions
- Commonly used in the pharmaceutical industry for validating drug purity
- Proteins are usually denatured after passage through HPLC

9.10 PROTEIN ANALYSIS

Edman degradation

- Used to determine the sequence of amino acids in a peptide, even of protein fixed in an SDS-PAGE gel then immobilized on a polyvinylidene fluoride (PVDF) membrane
- Highly sensitive—can sequence as little as 10–100 ng of protein
- Carried out in an automated analyser that conducts all steps:
    i. coupling thiocyanate group to the N-terminal amino acid of a peptide
    ii. cleavage to release the modified N-terminal amino acid
    iii. conversion of the modified amino acid to a stable derivative by heating in acid
    iv. identification of the stable amino acid derivative by reverse phase HPLC
- The cycle then repeats to identify the next amino acid, and so on
- Many proteins are 'blocked' at the N-terminus by, for example, formyl or acetyl groups so must be cleaved by chemical (CNBr) or enzymatic (*trypsin*) degradation to individual peptides prior to Edman degradation
- Now superseded by proteomics liquid chromotography–mass spectrometry

Measuring protein concentrations

- Absorption of UV light at 280 nm A_{280} can be used:
    i. an A_{280} of 1.0 in a 1 cm path length represents a protein concentration of ~1 mg/ml
    ii. if nucleic acids are also present, it is necessary to apply a correction factor: protein concentration (in mg/ml) is given by $1.55 \times A_{280} - 0.76 \times A_{260}$
- UV absorption is used routinely in detectors attached to chromatography systems
- Bradford assay: commonly used, relies on binding of Coomassie brilliant blue to protein under acidic conditions:
    i. light absorbance is measured at 595 nm (A_{595}) compared with a calibration curve, e.g. of different concentrations of bovine serum albumin (BSA)
    ii. it is affected by the content of basic amino acids (lys and arg), and by detergents
    iii. some proteins will precipitate under the acidic conditions, so cannot be measured in this assay

Protein gels

- Protein samples are heated to ~95°C for 3–5 minutes with:
  - i. ionic detergent sodium dodecyl sulphate (SDS)—results in denaturation of the protein and confers an overall negative charge on the protein that is proportional to the protein's size
  - ii. reducing agent (e.g. DTT or β-mercaptoethanol)—results in breakage of any disulphide bridges ensuring that the protein is monomeric and unfolded
- Samples are loaded on polyacrylamide gels containing SDS and separated by passing an electric current through the gel (SDS-PAGE)
- Proteins migrate towards the positive pole (as they are negatively charged by coating with SDS) according to mass—size can be determined by running a set of proteins of known size in parallel (molecular weight marker)
- Proteins can then be visualized in the gel by staining with Coomassie brilliant blue or silver or transferred to a nylon or nitrocellulose membrane for immunoblotting

Two-dimensional IEF

- Isoelectric focusing (IEF) can be used to separate proteins according to their **isoelectric point (pI)** in a first dimension gel (strip or tube) containing ampholytes
- The first dimension gel is then laid across the top of a standard SDS-PAGE gel and the proteins separated according to mass
- This technique allows discrimination between proteins of the same mass that have different pI, and even between different post-translational modifications of the same protein such as phosphorylation
- Comparative 2D gels can be used to determine if protein levels change between different experimental conditions, e.g. upon treating cells with a drug
- 2D-IEF gels formed the original basis of proteomic analysis

Proteomics

This modern technique analyses all the proteins present in a collection of proteins (e.g. whole cell lysate) and can provide either qualitative or quantitative comparisons of protein levels under related experimental conditions, e.g. to determine the effect of drug treatment on the protein profile of a cell.

- Samples are prepared then separated on 2D electrophoresis (see 2D-IEF above)
- Gels are stained:
  - i. current methods allow protein detection up to picomole (10^{-12}) and femtomole levels (10^{-15})
  - ii. new techniques under commercial development dramatically improve sensitivity, allowing protein detection down to zeptomole level (10^{-21}: less than 600 molecules). The level of detection is greater than 10,000 times more sensitive than existing methods in general use

- Protein patterns on 2D gels are analysed using powerful image analysis techniques and software. Proteome maps are compared to detect proteins that are up- or downregulated in different experimental conditions (e.g. cancer vs. normal, cells treated ± drug)
- Proteins are excised from 2D gel spots for identification and full characterization (this can be done by robots):
    i. proteins are then digested (e.g. by trypsin) and analysed by mass spectrometry
    ii. advanced peptide mass fingerprinting and mass spectrometry methods allow generation of a protein 'tag' that can be used to interrogate databases
- Proteome data generated are entered into extensive protein databases and analysed using bioinformatics technology
- Newer proteomics platforms avoid the need for 2D gels and can conduct mass spectrometry on large, mixed protein populations by mass spectrometry
- Differential labelling of test and reference samples allows both identification of proteins that differ between two experimental conditions and quantification of the differences, e.g. SILAC labelling (stable isotope labelling of amino acids in cells)

   see also 9.12 mass spectrometry (p. 265)

Looking for extra marks?

Variations of proteomics allow analysis of post-translational modifications of proteins including glycosylation and phosphorylation. This is extremely useful as such modifications can affect the protein's activity.

Check your understanding

How can cloning protein-coding genes help us isolate the proteins? (*Hint: consider the use of 'tags' on recombinant proteins and control of gene expression levels.*)

Describe what is meant by electrophoresis. How can this technique be applied in the characterization of (a) proteins and (b) DNA? (*Hint: remember the charges on DNA, and on proteins after treatment with SDS. Think about the type of gels used— acrylamide and agarose—and why each is chosen. For extra marks in part (a), you could discuss 2D-IEF.*)

9.11 IMMUNOLOGICAL TECHNIQUES

Immunological techniques exploit the high specificity and affinity of antibodies for their cognate antigen. They can be used simply to analyse the presence of a protein, examine post-translational modifications, probe protein–protein interactions, and interactions of protein with other macromolecules (e.g. DNA). Antibodies can also be used to provide quantitative determination of protein concentrations using competition assays.

Antibodies

- Antibodies are proteins produced by the immune system that have high affinity and high specificity for a particular antigen
- The majority of antibodies used in biochemical techniques are IgG molecules with two heavy and two light chains, linked by disulphide bonds
- IgGs have two antigen binding sites per molecule formed by the variable Fv region within the Fab domain
- All antibodies of a particular species of organism (e.g. mice) share a common region (the Fc domain), which can be recognized either by antibodies from a different species (e.g. rabbit), or bound directly to protein A, produced by *Staphylococcus aureus*
- The Fc domain can be swapped using molecular biology techniques to prevent the antibody being recognized as foreign, e.g. using the human Fc domain on a mouse monoclonal antibody to allow its use as a drug in humans
- **Polyclonal antibodies** recognize the same antigen but are produced by different B cell clones, and may recognize different epitopes on the antigen:
    i. generated by immunizing an experimental animal (mouse, rabbit, sheep, etc.) with purified antigen (often recombinant purified protein—called the immunogen) and by further stimulating an immune response using an adjuvant
    ii. after a few weeks, antibodies against the immunogen are produced
    iii. to increase yield and specificity, further 'booster' injections of the immunogen are given
    iv. serum is collected, which contains high levels of the antibody of choice: this can be further purified by collecting the total IgG fraction (e.g. by binding to protein A) or by affinity purification using the immunogen to capture the antibody from solution
- **Monoclonal antibodies** are produced from a single B cell clone and so all recognize the same epitope on the antigen; they are secreted by immortal cells grown in tissue culture and so provide a theoretically infinite supply of antibody:
    i. produced by immunizing an animal (usually mouse) as above
    ii. spleen is harvested and dissociated into single cells (these produce antibodies)
    iii. spleen cells are fused with myeloma cells (B-cell-lineage cancer cells that do not make their own antibodies but that are immortal)
    iv. fused cells are selected for using HAT (hypoxanthine, aminopterin, and thymidine) which kills unfused myeloma cells—the spleen cells die unless fused as they are mortal
    v. fused cells are plated out in tissue culture dishes at a density of one cell per well of a multiwell dish
    vi. secretion of the required antibody is tested

Western blotting (immunoblotting)

A technique to identify specific proteins on a gel (SDS-PAGE or 2D gel) by their binding to an antibody with known specificity.

- Proteins are separated on SDS-PAGE or 2D PAGE, then electrophoretically transferred onto nitrocellulose or PVDF/nylon membrane in the presence of buffer and methanol (which fixes the proteins onto the membrane)
- Non-specific sites on the membrane are blocked by incubating it with BSA or milk proteins
- The blocked blot is then incubated with a primary antibody specific to the protein of interest: this antibody will bind to its cognate antigen on the blot (if present)
- After washing, the blot is incubated with an enzyme-conjugated secondary antibody from a different species that recognizes the Fc portion of the primary antibody
- Binding of this secondary antibody is visualized by virtue of the chemical reaction catalysed by the conjugated enzyme, e.g.
    - i. colour change (*alkaline phosphatase* enzyme)
    - ii. light emission (enhanced chemiluminescence) and exposure to X-ray film (*horseradish peroxidase, HRP*) (see Figure 9.18)

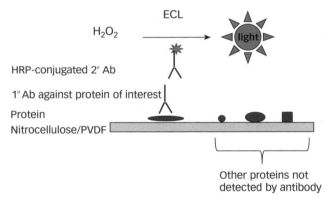

Figure 9.18 Western blot detection of proteins separated by SDS-PAGE and transferred to a nitrocellulose or PVDF membrane. Non-specific binding is blocked by saturation of the membrane with non-specific protein (e.g. milk proteins)—not shown. Detection method shown here is enhanced chemiluminescence (ECL). 1° Ab = primary antibody; 2° Ab = secondary antibody.

Immunoprecipitation (IP)

Used to isolate protein of interest (and any associated protein partners) from a complex mix of proteins.

- Antibodies specific to protein of interest are incubated with a complex mix of proteins, e.g. cell lysate
- Antibody molecules are immobilized via their Fc portion, e.g. using protein A attached to large Sepharose beads

- The beads are collected by gentle centrifugation, precipitating also the bound protein (and any other proteins with which it is associated)
- After washing, proteins can be released from the beads by boiling in SDS with reducing agents, and proteins separated on SDS-PAGE (NB this treatment breaks apart the heavy and light chains of the antibody molecules as well as dissociating the protein of interest from the antibody)
- Has been used to identify components of, for example, the Fanconi anaemia complex—antibodies against one component co-immunoprecipitate other components of the complex

Chromatin immunoprecipitation (ChIP)

This powerful technique enables identification of DNA sequences bound by specific proteins. It can even be used to differentiate between modified versions of the same protein bound to DNA (e.g. epigenetic analysis).

- Nuclear extracts from cells are treated with formaldehyde to stabilize (cross-link) interactions between proteins and the DNA to which they are bound
- The DNA is exposed to a nuclease (e.g. *micrococcal nuclease*), or physically sheared by sonication
- An antibody specific to the protein of interest is added, and used to specifically precipitate the protein and any DNA to which the protein is bound
- The protein cross-linking is reversed by heating
- The DNA is amplified by PCR either using primers against known regions of DNA to determine if that region was associated with the protein, or using 'adaptors' which are short oligonucleotides annealed to the ends of the precipitated DNA against which specific PCR primers can be designed

  ➡ *see 9.1 DNA manipulation and analysis (p. 227) for PCR*
- The amplified DNA is sequenced to determine the region to which the protein was bound
- For steps in ChIP see Figure 9.19
- It is now possible to sequence the precipitated DNA directly using next generation solid state sequencing methods—this is known as ChIP-Seq

Radioimmunoassay (RIA)

A method for determining protein concentration in a complex mix, using competition for binding to an antibody specific to the protein of interest. It uses radiolabelled antigen as a tracer; unlabelled antigen competes with the labelled antigen for the antibody binding site.

- Requires a purified sample of the protein of interest at known concentration, to establish a calibration curve
- A subset of the purified protein (antigen) is radiolabeled, e.g. [125]I

Immunological techniques

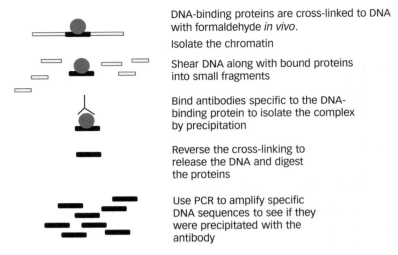

DNA-binding proteins are cross-linked to DNA with formaldehyde *in vivo*.
Isolate the chromatin

Shear DNA along with bound proteins into small fragments

Bind antibodies specific to the DNA-binding protein to isolate the complex by precipitation

Reverse the cross-linking to release the DNA and digest the proteins

Use PCR to amplify specific DNA sequences to see if they were precipitated with the antibody

Figure 9.19 Chromatin immunoprecipitation (ChIP).

- A fixed amount of radiolabelled antigen (Ag) is incubated with its specific antibody (Ab), and increasing amounts of unlabelled purified Ag of known concentration are added in sequential samples (see Figure 9.20)
- Ab is collected by binding to protein A-agarose beads and centrifugation (with washing steps)
- Amount of radioactivity precipitated is measured in a scintillation counter
- As the concentration of unlabelled Ag increases, more binds to the Ab, the labelled Ag is displaced and radioactive signal decreases, i.e. there is an inverse relationship between the amount of unlabelled Ag bound and the amount of radioactivity precipitated—this allows the derivation of a calibration curve
- Increasing amounts of the unknown sample are then incubated with a fixed amount of radiolabelled Ag and Ab, and the Ab (with bound Ag) is precipitated and radioactivity measured.
- The amount of Ag present in the unknown sample can be determined by comparison with the calibration curve
- This is a useful technique for determining, for example, hormone concentrations in patient blood samples

ELISA (enzyme-linked immunosorbent assay)

This assay can detect the presence of a specific protein in a complex mix of proteins (e.g. cell lysate) and competitive ELISA can provide quantitative measurement of protein concentrations. ELISA can also be used to detect interactions between different proteins.

- Antibody is immobilized onto wells of a multiwell plate; non-specific binding sites are blocked—as in Western blotting
  - (➔) *see 9.11 Immunological techniques (p. 256) for details of Western blotting*

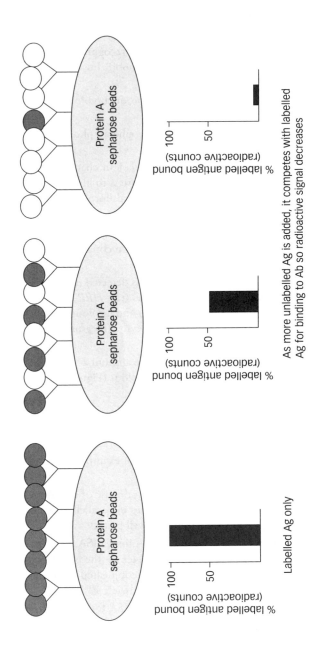

Figure 9.20 Radioimmunoassay (RIA) is based on competition for antibody binding. Radiolabelled antigen in solution is captured on antibody (which is immobilized by binding, e.g. to protein A–sepharose beads). Unlabelled antigen, e.g. from a patient sample, competes with the radio-labelled antigen for binding to the antibody. Higher unlabelled [Ag] results in less radiolabel precipitated—it is an inverse relationship.

Immunological techniques

- Antigen is added; any unbound antigen is washed off
- Bound antigen is detected:
    i. directly, if Ag was labelled (e.g. biotinylated Ag is detected by addition of enzyme-conjugated streptavidin)
    ii. indirectly, by adding a second antibody that recognizes a different epitope on the antigen than that bound by the immobilizing antibody; this second antibody can be enzyme-conjugated
- Commonly used enzymes produce a detectable product (colour, luminescence, etc.). They include:
    i. *alkaline phosphatase* (colour change of nitrophenol phosphate, colourless to yellow)
    ii. *HRP (horseradish peroxidase)*—detected by colour change (e.g. tetramethylbenzidine converted from colourless to blue product) or chemiluminescence, dependent on free radicals formed on breakdown of H_2O_2 by *HRP*
- Competitive ELISA can be used to measure concentrations of antigen:
    i. increasing amounts of unlabelled Ag of known concentration are added to wells coated with Ab
    ii. after washing off unbound Ag, bound Ag is detected using an enzyme-conjugated molecule that binds to the label (e.g. *HRP*-conjugated streptavidin which binds tightly to biotin) and a colour change or luminescence reaction catalysed by the enzyme (measurements taken using a spectrophotometer plate reader)
    iii. there is an inverse relationship between the amount of unlabelled Ag bound to the Ab and the intensity of the colour (Figure 9.21)—this allows the derivation of a calibration curve
    iv. the concentration of Ag in an unknown sample can be determined by carrying out further competition assays and interpolating from the calibration curve
    v. this is a useful technique for determining, for example, hormone concentrations in patient blood samples
    vi. it has the benefit over RIA of not requiring the use of radioactivity, and the possibility of conducting multiple replicates in parallel, or of testing many more samples at one time (e.g. using 96-well plates)
- Protein–protein interactions can be determined using ELISA, e.g. by immobilizing Ab against one protein of the complex on the plate, incubating with the protein mix, then detecting using an enzyme-conjugated Ab against the other protein of the complex—this is a variation of 'sandwich' ELISA
- Colour change pregnancy tests are based on the ELISA format—competition between human chorionic gonadotrophin (hCG) in urine and immobilized hCG in a paper strip releases enzyme-conjugated anti-hCG antibody, which moves by capillary action to the detection window where it forms a sandwich with a second immobilized anti-hCG via the hCG in the urine

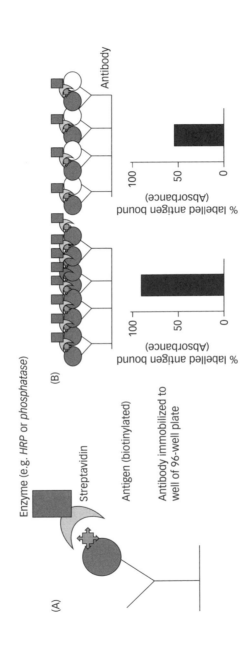

Figure 9.21 ELISA (enzyme-linked immunosorbent assay). There are many different possible configurations. (A) A simple capture assay using antigen (Ag) labelled with biotin and detected using enzyme-conjugated streptavidin. (B) Competitive ELISA for quantification: unlabelled antigen competes with labelled antigen, so as more unlabelled Ag is added, the signal decreases, i.e. there is an inverse relationship between [Ag] and signal detected.

Immunohistochemistry (IHC) and immunofluorescence (IF)

- Antibody specific to a particular protein of interest (the antigen) is incubated with tissue/cell sample fixed and immobilized on a microscope slide
- After washing, the antibody is detected by incubating with a secondary antibody which binds to the Fc protein of the primary Ab
- The secondary antibody is conjugated to an enzyme which catalyses a colour change reaction (immunohistochemistry), or a fluorophore which emits light on excitation with a specific wavelength (immunofluorescence). Using a secondary Ab also amplifies the signal
- The signal is viewed using a microscope (with specific wavelength filters for immunofluorescence)
- This technique is useful for determining the localization of proteins in cells or tissues, and relative levels (e.g. in cancer tissue samples)

9.12 BIOPHYSICAL TECHNIQUES

Circular dichroism (CD)

A relatively quick method for analysing aspects of protein structure in solution.
- Depends on differential absorption of left and right circularly polarized light
- UV-CD is used to assess degree of secondary structure of proteins, i.e. alpha-helix, beta-sheet, beta-turn, and random coil (it cannot assign where these motifs are within a molecule):
    i.   effect of mutation on overall protein conformation can be determined by comparing CD spectra of the wild-type and mutant proteins
    ii. protein stability can be determined under conditions of increasing temperature, pH changes, etc.
    iii. effect of ligand binding on protein structure and stability can be determined by looking at changes in CD spectra ± ligand
- Near-UV CD (>250 nm) can be used to assess tertiary structure because of effects on phe, tyr, cys, and trp amino acids—useful especially in providing information on prosthetic groups such as **haem** in haemoglobin and cytochrome c
- Visible CD is used to study interactions between proteins and metals since CD spectra in the visible range are only produced when metals are in a chiral environment
- Limitations: CD does not give atomic level resolution and is not readily applied to intrinsic membrane proteins

X-ray crystallography

This powerful technique allows determination of the structure of biological molecules including details such as bond lengths and angles. It is only suitable for molecules that can be crystallized. There are three major steps: (i) crystallization; (ii) exposure to an intense beam of X rays and collection of patterns of reflections; and (iii) mathematical computation (Fourier transformation) to work out the arrangement of atoms in the molecule.

1. Crystallization

- The molecule to be studied is crystallized (by a process of trial and error, screening for optimal conditions) by promoting first nucleation then growth of a crystal
- The aim is to achieve a single large crystal (> 0.1 mm in all dimensions) rather than multiple small crystals
- The solubility of the component molecules is gradually lowered, e.g. by altering pH, changing the dielectric constant of the solution, or by supersaturating the solution (lowering temperature can promote supersaturation and aid crystallization)
- Requires very high concentrations of very highly purified protein (e.g. milligram amounts at ~8–10 mg/ml)
- Robotic methods of pipetting allow screening trials using only 100 nl per sample in multiwell plates, where each well contains slight differences in buffer conditions
- Detergents can inhibit crystallization, so it is difficult to obtain crystals of integral membrane proteins (e.g. ion channels) because detergents are required to release the proteins from the biological membrane
- Denatured (unfolded) proteins cannot be crystallized

Looking for extra marks?

Impurities, detergents, and unfolding of proteins all interfere with crystallization. This is related to the forces packing protein molecules together in the crystal and the requirement for repeated, identical interactions. Consider specifically the sorts of forces involved in packing and how these factors may interfere with them.

2. Single crystal X-ray diffraction

- The aim is to determine the density of electrons (and hence the positions of the various atoms) throughout the crystal
- The crystal is mounted on a rotating platform called a goniometer, then is flash frozen in liquid nitrogen to reduce radiation damage to the crystal and to decrease thermal motion which affects resolution
- It is bombarded with X-rays of a single wavelength, e.g. from a synchrotron source (typically ~0.1 nm)
- The X-rays are diffracted by the crystal lattice producing a pattern of regularly spaced spots called reflections, as they are caused by reflection of X rays from one set of evenly spaced planes within the crystal
- Reflections are detected using a highly sensitive CCD (previously collected on X-ray film)—they give information on both the strength and angle of beam diffracted (and about atomic oscillations within the crystal)

Biophysical techniques

- The crystal is then rotated step-by-step through 180° (less for very regular crystals) and a series of reflections is collected at each rotation (tens of thousands of reflections are typically collected)
- This is elastic scattering, i.e. the scattered X-rays have the same energy and wavelength as the X-rays hitting the crystal, but their direction is altered
- For very pure crystals, chemical bond lengths and angles can be determined very precisely (few thousandths of an Ångström and few tenths of a degree, respectively)
- Variations:
  i. Laue scattering uses X-rays of multiple wavelengths, useful in studying very rapid events—time-resolved crystallography
  ii. Laue back reflection—if the sample is too thick for X-rays to pass through
  iii. X-ray fibre diffraction—used by Rosalind Franklin in determining the structure of DNA
  iv. phasing, e.g. multi-wavelength anomalous dispersion (MAD) uses X-rays of different wavelengths

3. Fourier transformation

- The 2D data collected are converted to 3D models of electron density using Fourier transformation
- Reflections are indexed to identify which reflection represents a particular position in space—this also allows determination of the symmetry of the crystal
- Images from different angles are merged to identify reflections present in more than one orientation, and these are used to scale the intensity of the images
- The intensity of each reflection is related to the structure factor amplitude
- The phase of the wave must be determined in order to calculate the electron density map—this is done directly (for small molecules) by molecular replacement (if a related structure is known), by introducing electron dense atoms directly into the crystal by soaking, or more commonly by anomalous scattering by selenium incorporated via seleno-methionine in methionine-auxotrophic bacteria during recombinant protein expression
- An initial model is then constructed and refined (using the known protein sequence and known physical properties of the amino acids) to fit the diffraction data
- Further models are then produced until the model and diffraction data correlate as closely as possible

Nuclear magnetic resonance (NMR) spectroscopy

A powerful technique that provides structural information on biological molecules including their topology, structure, and dynamics. It can be used on proteins that cannot be crystallized as it analyses molecules in solution, but is limited to proteins of <50 kDa.

- Proteins must be highly purified and present at 0.1–3 mM, though only small volumes are required (300–600 µl), cf. much larger amounts are needed for X-ray crystallography
- Isotopic labelling (e.g. with ^{13}C or ^{15}N) is preferred—can be achieved by expression of recombinant protein *in vitro*
- Samples are placed in a thin-walled glass tube and exposed to pulses of radiofrequency electromagnetic radiation in a strong magnetic field
- Each ^1H, ^{13}C, or ^{15}N nucleus within a molecule has a distinct chemical environment and so will produce a distinct chemical shift
- HSQC = 2D heteronuclear single quantum correlation spectrum for ^{15}N-labelled proteins: each amino acid (except proline) gives one signal for the amide–hydrogen bond (trp and other N-side chain amino acids can give more than one signal)
- The ^{15}N-HSQC spectrum provides a protein's 'fingerprint'
- COSY = correlation spectroscopy, including total correlation spectroscopy (TOCSY) and nuclear Overhauser effect spectroscopy (NOESY)—can be used on unlabelled proteins
- In COSY and TOCSY, bonding patterns are measured: each amino acid provides a characteristic pattern and can be identified. NOESY is affected by through-space interactions (distance between nuclei)
- Signals can be used to calculate distances between nuclei according to peak intensity, with chemical shifts used to calculate angle restraints
- The restraints determined by experiment are used in computer programs to calculate the structure based on known properties of proteins, including bond lengths and angles, attempting to minimize the energy, i.e. to produce a structure that fits experimental observations and is the most stable configuration
- In large proteins, there is significant overlap between peaks so better to use ^{13}C and ^{15}N labelled proteins
- NMR can also be used to study dynamics by measuring relaxation times, e.g. for ^{15}N, backbone motions between 10 picoseconds and 10 nanoseconds can be determined
- ^{13}C and deuterium relaxation methods allow analysis of motion of side chains

Mass spectrometry (MS)

A method for measuring particle mass and determining composition of molecules, e.g. peptides. Molecules are ionized to generate charged particles and their mass-to-charge ratio is determined. Mass spectrometry can be used both qualitatively and quantitatively, and now forms the basis of proteomic techniques for identifying proteins by a peptide signature.

- The sample is vaporized in a mass spectrometry (MS) instrument
- For small molecules (e.g. drugs), the molecules are ionized, e.g. by bombarding with electrons

Biophysical techniques

- Protein molecules are typically ionized in solution (but are non-volatile). Protein ions are introduced into the vapour phase directly, e.g. by evaporating the solvent in electrospray ionization (ESI), or matrix-assisted laser desorption/ionization (MALDI)
- Charged particles are accelerated in an electric field (and steered by a magnetic field) according to mass-to-charge ratio m/z. Detection is either by their position in the field (quadrupole detection) or the length of time taken to reach the detector (time of flight (TOF) analysis)
- A mass chromatogram is derived (intensity on y-axis, mass/charge on x-axis)
- Can be combined with chromatographic techniques including gas chromatography–mass spectrometry (GC-MS) and liquid chromatography–mass spectrometry (LC-MS)
- Whole proteins can be analysed by either ESI-MS or MALDI-MS. The mass can be determined very precisely (±1 mass unit), which is often sufficient to identify that protein uniquely in a genome database
- In proteomics, proteins are digested by, for example, *trypsin*; then the peptide fragments are identified by MS to give a peptide mass fingerprint that can be used to identify the original protein
- Glycans can be readily studied by MALDI-MS, or ESI-MS for smaller glycans
- Measurement of drug metabolism in body fluids is achieved by LC-MS to determine pharmacokinetics

Surface plasmon resonance (SPR)

This technique uses very small amounts of protein and can analyse protein–protein and protein–ligand interactions in real-time, providing kinetic information including K_d and order of reaction.

- Protein of interest (target) is immobilized on a gold-coated glass slide
- When light is shone on this thin metal film, a fraction of the light energy incident at a sharply defined angle can interact with delocalized electrons in the metal film (plasmon), decreasing the *reflected* light intensity
- If another protein binds to the immobilized target, the local refractive index changes, leading to a change in SPR angle, which can be monitored in real-time by detecting changes in the intensity of the reflected light, producing a sensorgram (see Figure 9.22)
- The rate of change of the SPR signal yields apparent rate constants for the association and dissociation phases of the reaction and hence apparent equilibrium constant (affinity)
- The size of the change in SPR signal is directly proportional to the mass being immobilized, and can thus be interpreted crudely in terms of the stoichiometry of the interaction
- Signals are easily obtained from sub-microgram quantities of material

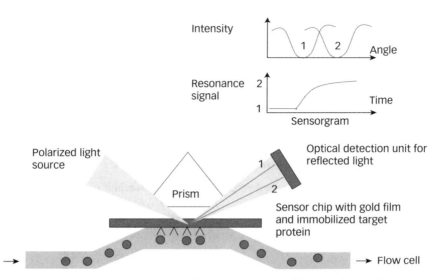

Figure 9.22 Surface plasmon resonance (SPR). The protein of interest (target) is immobilized on a gold-coated glass slide. Incident light interacts with delocalized electrons in the metal film (plasmons), causing a decrease in reflected light intensity (1). Local refractive index changes if another protein binds to the immobilized target (2). Changes in SPR angle are monitored in real-time by detecting changes in the intensity of the reflected light—produces a sensorgram.

Single particle electron microscopy

A method to obtain structural information at nanometre to atomic resolution using transmission electron microscopy.

- Highly purified homogeneous protein is placed on an EM grid and negatively stained
- Images are obtained using transmission electron microscopy (TEM)—uses low dose electrons to avoid damage to biological samples
- Necessary to group images into different classes as the protein will be present in different orientations on the grid (and may exist in different conformations)
- Computerized image processing combines multiple (10^3 to 10^4) digitized images (= averaging) to improve the signal:noise ratio—generates 2D images of the protein/particle (e.g. virus)
- Image filtering can be used to improve the resolution
- 3D images can be generated by combining 2D projections taken at different viewing angles:
    i. random conical tilt
    ii. orthogonal tilt
- Cryo-EM is a variation on this that does not use stains:

  i. protein samples are embedded in <u>vitreous</u> ice to preserve a physiological environment, with no need for fixation or staining

 ii. samples are viewed by TEM at cryo-temperatures (liquid nitrogen or liquid helium)

iii. resolution can be low so image data are combined with data from X-ray crystallography

- CryoEM tomography (CET) allows 3D reconstruction of 2D cryo-EM images

 Check your understanding

Why is it difficult to determine the 3D structure of a protein if its sequence is not known? (*Hint: how is the protein isolated in the first place? If it has been cloned, then it is much easier to produce large amounts for structural analysis. Also remember the modelling and reconstructions that are required in refinement, e.g. in X-ray crystallography.*)

How might circular dichroism be used to study protein unfolding? (*Hint: what features of protein structure are observed by CD? What happens to these when the protein unfolds?*)

Glossary

26S proteasome Very large protein complex (~2000 kDa) involved in breaking down proteins; comprises one 20S core particle and 2 × 19S cap structures; degrades unwanted or damaged proteins, e.g. those marked by the ligation of ubiquitin.

acetate Anion derivative of acetic acid, forming salts or esters (CH_3COO^-).

acetoacetate Component of a ketone body, water-soluble form of transferring the carbon atoms from fatty acids for metabolism. See also acetone and β-hydroxybutyrate.

acetone Component of a ketone body, water-soluble form of transferring the carbon atoms from fatty acids for metabolism. See also acetoacetate and β-hydroxybutyrate.

acetyl CoA Breakdown product from glycolysis, fatty acid, and some amino acid metabolism that can enter TCA cycle to generate energy, or be used for biosynthesis.

acetyl CoA synthetase Enzyme that generates acetyl CoA.

acid Proton donor (cf. base).

active site Amino acid residues on enzyme surface that contact substrate and/or transition state. Often in cleft in protein.

acute Active over short periods of time.

acyl carrier protein Carries acyl groups between enzymes of fatty acid synthase.

adenine Purine base found in DNA and RNA and other compounds; e.g. ATP, NADH.

adenovirus DNA viruses often used to introduce recombinant DNA into mammalian cells or whole organisms.

adenyl cyclase Membrane-bound enzyme that converts ATP to cAMP; aka adenylyl cyclase.

adipocyte Cell which stores triacylglycerols in fat droplets.

adrenaline Hormone used to signal 'fight or flight' in animals causing up-regulation of catabolism and inhibition of anabolism. Also known as epinephrine.

alcohol dehydrogenase Enzyme that reduces ethanal to ethanol regenerating oxidized NAD^+. Used in ethanol production in yeast/bacteria, or in ethanol metabolism in mammalian liver.

aldehyde dehydrogenase Enzyme required for metabolism of alcohol in mammalian liver.

allosteric Means of regulating enzyme activity by binding a small molecule at a site distinct from the active site.

allosteric site Binding site for a regulatory molecule that is distinct from the active site.

alpha-helix Secondary structure of proteins stabilized by hydrogen bonds between peptide bonds four residues apart on same peptide chain (intramolecular).

amino acid Building block of protein with one amino group and one carboxylic acid group separated by a central carbon. Variable R side chain conferring specific chemical properties.

aminoacyl-tRNA Amino acid linked to a tRNA molecule for incorporation into growing peptide chain in translation.

amphipathic Molecule with hydrophilic region and hydrophobic regions.

anabolism Biosynthesis of complex molecules from simple building blocks.

apoptosis Programmed cell death or cell suicide.

Glossary

assay Experimental procedure for determining amount of material present or rate of a reaction.

ATP synthase Enzyme complex found in inner mitochondrial membrane that uses H^+ gradient to generate ATP.

attenuation Mechanism used to regulate gene expression by inducing premature termination of transcription by *RNA polymerase*. Used by operons encoding amino acid synthetic pathways in bacteria, e.g. trp.

attraction A force between molecules which promotes their interaction.

autophosphorylation Enzyme activity that results in phosphorylation of the same enzyme molecule, e.g. in *receptor tyrosine kinases*.

autotroph Organism able to make organic material when provided only with small inorganic molecules (CO_2, N_2).

bacterial artificial chromosomes (BACs) Circular artificial chromosomes that can replicate and segregate correctly during the cell division process in bacteria. Used to clone large fragments of DNA.

bacteriophage Virus infecting bacteria, e.g. phage lambda.

baculovirus Virus infecting insect cells. Vectors based on baculovirus are used to express large amounts of protein in insect cells.

base Proton acceptor (cf. acid). DNA/RNA contains nucleotide (heterocyclic) bases, which may be referred to simply as 'bases' in context.

base tautomerization H atoms on the bases transiently move position and allow the formation of non-standard base pairs that fit in the double helix. This can lead to mutations, e.g. the imino ($=NH$ rather than $-NH_2$) form of A can base pair with C.

beta-oxidation Pathway by which fatty acids are broken down, two carbons at a time, to acetyl CoA which can be used to generate energy or in biosynthesis.

beta-pleated sheet (β-sheet) Secondary structure of proteins stabilized by hydrogen bonds between peptide bonds on adjacent chains (inter- or intramolecular).

bidirectional In two directions from a central point, e.g. DNA replication.

bilayer Usually used to describe two layers of phospholipids, with polar headgroups facing outwards and hydrophobic fatty acids buried inside, forming a biological membrane.

branched chain amino acids Valine, leucine, and isoleucine, which have a branched hydrocarbon side chain.

C4 pathway Ancillary carbon fixation pathway in some plants, using spatial separation of carbon fixation from Calvin cycle.

Calvin cycle Cycle for regeneration of ribulose 1,5-bisphosphate after CO_2 fixation. Involves sugar interconversions similar to pentose phosphate pathway.

CAM pathway crassulacean acid metabolism ancillary carbon fixation pathway in some plants, using temporal separation of carbon fixation from Calvin cycle.

cAMP Cyclic AMP—adenosine 3′,5′-cyclic monophosphate. A second messenger formed in cells from ATP by *adenylyl cyclase*.

carnitine acyl transferase *Transferase* used to transfer fatty acids into mitochondria. One carrier is present on each of the mitochondrial membranes.

cascade Series of reactions in which each step generates more molecules than the previous step. Leads to amplification of signals.

catabolism Breakdown of molecules to generate energy (or building blocks of other molecules).

catalytic efficiency Fraction of enzyme–substrate collisions that result in reaction.

centromere Region of DNA where two sister chromatids come into close contact which attaches to spindle and facilitates chromosome segregation during division.

chemical cleavage Cleavage of a bond using chemicals rather than enzymes.

chiral Lacking an internal plane of symmetry and therefore having a non-superimposable mirror image.

chlorophyll Heterocyclic conjugated ring molecule, similar to haem, except containing chelated Mg^{2+}. Absorbs light in photosynthesis.

chloroplast Organelle found in plants where photosynthesis takes place.

cholesterol Lipid found in eukaryotes to regulate membrane fluidity.

chromatin Combination of DNA and proteins found in the nucleus of a eukaryotic cell.

chromosome Single piece of double-stranded DNA found in a cell. Must contain sequences to allow DNA replication, segregation during division (centromere in eukaryotes) and, if linear, telomeres to protect the ends.

chronic Active over long periods of time.

chylomicron Lipoprotein particle for transport of lipid in blood.

citric acid cycle Cycle of chemical reactions by which reducing power is extracted from acetyl CoA within mitochondria. Also known as TCA or Krebs cycle.

codon Three adjacent nucleotides in a nucleic acid that code for a particular amino acid.

cofactor Non-amino acid molecule/ion bound to a protein which participates in its function (e.g. Zn(II), biotin)

committed step Reaction in a pathway which commits a substrate to a particular eventual endpoint.

compartmentalization Segregation of functions between different organelles within a eukaryotic cell. Sometimes used to describe segregation between different tissues.

competitive inhibition Reversible inhibition that can be overcome by increased substrate binding.

complementary base pairing Two strands of DNA (or RNA) joined by H-bonds between bases. A always pairs with T or U in mRNA), and G always pairs with C.

complementation assay Identification of a mutated gene by introduction of a piece of DNA from a non-mutated cell that can revert the phenotype of a mutated cell.

consensus sequence Sequence derived by comparing multiple similar sequences to determine the most common. Usually applied to short sequences of DNA that are bound by specific proteins, e.g. transcription factors.

co-operativity In enzymes, situation where the first molecule bound stimulates the binding of further identical molecules, i.e. binding becomes easier after the first molecule is bound.

copy number Number of copies of a molecule found in a cell.

Cori cycle Inter-organ cycle in which lactate (released by muscle) is converted back to glucose (in the liver) and exported back to muscle.

cosmid Cloning vector, based on plasmid, which allows insertion of large fragments (up to 40 kb) of foreign DNA. Contain cos sites from bacteriophage lambda to allow packaging into phage.

covalent bond Bond formed by sharing electrons between two atoms.

cyanobacteria Only bacteria capable of oxygenic photosynthesis (previously called blue-green algae).

cytochrome Electron carrier containing haem used in electron transport chain in mitochondria and photosynthesis in chloroplasts.

cytosine Pyrimidine base found in DNA and RNA.

cytoskeleton Network of filaments found in the cytoplasm of a eukaryotic cell for transport, strength, etc. Made up of microtubules, intermediate filaments and microfilaments.

Glossary

cytosol Soluble portion of a cell outside organelles.

degenerate primers Mixture of DNA oligonucleotide primers containing more than one nucleotide sequence encoding the same amino acids.

deoxyribonucleic acid A polymer of deoxyribonucleotides linked by phosphodiester bonds. Exists as double helix and forms the genetic material of most living organisms. Also known as DNA.

deoxyribonucleoside Nitrogenous bases linked to the C1 carbon of 2′ deoxyribose by a β-*N*-glycosidic bond, i.e. base + sugar.

deoxyribonucleotide deoxyribonucleoside with the 5′ carbon of deoxyribose joined by an ester linkage to a phosphate group (dNMP) with or without additional phosphates (dNDP, dNTP), i.e. base + sugar + phosphate(s).

deoxyribonucleotide triphosphate Nucleoside with the 5′ carbon of deoxyribose bound by an ester linkage to three phosphate groups (dNTP).

diabetes Disease where tissues are unable to use blood glucose effectively.

DNA Deoxyribonucleic acid, a polymer of deoxyribonucleotides linked by phosphodiester bonds. Exists as double helix and forms the genetic material of most living organisms.

DNA double helix Two antiparallel strands of DNA held together by hydrogen bonds between base pairs, twisted into a helix.

DNA ligase Enzyme that can form a phosphodiester bond between 5′P and 3′OH of adjacent nucleotides on one strand of pre-existing DNA. Seals nicks in phosphodiester backbone.

DNA polymerase Enzyme that catalyses formation of a phosphodiester bond to generate new DNA. Always synthesizes DNA in a 5′ to 3′ direction. Requires a template and a primer as well as dNTPs.

DNA recombination Process by which a segment of DNA can move from one DNA molecule to another.

DNA repair Correction or replacement of damaged DNA.

DNA replication Duplication of the DNA genetic material in a cell.

domain Part of a protein that can fold and exist independently from the rest of the protein.

duplex Double-stranded molecule.

electrophile Reagent which accepts an electron pair to make a new bond in a chemical reaction (cf. *nucleophile*).

electrostatic bond Bond formed between two oppositely charged ions. Also known as an ionic bond.

elongation factor Protein that transiently associates with ribosome to facilitate translational elongation.

endocytosis Take up of material into cells by invagination of plasma membrane to form vesicles.

endoplasmic reticulum Membrane-bound organelle in eukaryotic cells which is the site of synthesis of lipid (smooth ER) and secreted and integral membrane proteins (rough ER).

endosymbiotic theory Theory that chloroplasts and mitochondria arose in eukaryotic cells by ingestion of bacteria.

enhancer Short sequence of DNA which binds a transcription factor to positively regulate transcription in eukaryotic cells. Can be upstream or downstream and a long distance from promoter.

enzyme Biological molecule that catalyses a chemical reaction.

enzyme assay *In vitro* measurement to detect rate and extent of enzyme activity.

(enzyme) saturation Situation where all active sites are occupied, such that increasing substrate concentration does not affect rate.

epinephrine Hormone used to signal 'fight or flight' in animals causing up-regulation of catabolism and inhibition of anabolism. Also known as adrenaline.

equilibrium reaction A reaction that can go in both directions. Direction *in vivo* is governed by the relative concentrations of reactants and products.

euchromatin Loosely packed form of chromatin, rich in genes.

exclusion bodies Particles of insoluble material found in the cytoplasm, e.g. a recombinant protein overexpressed in bacteria.

exclusion limit Upper limit of molecular weight above which all molecules will elute together in size exclusion chromatography.

exon Section of nucleic acid found in mature messenger RNA.

exonuclease Enzyme that degrades DNA from one end by breaking the phosphodiester bond joining the last nucleotide to the chain.

exponentially In terms of PCR, this means a doubling of product at each cycle (i.e. amount of product is 2^n where n = number of PCR cycles).

fatty acid Carboxylic acid with a long chain of CH_2 (typically C14, C16) used to store energy and synthesize phospholipids.

fatty acid synthase Complex of enzymes used to synthesize fatty acids.

feedback regulation Inhibition of a pathway due to increase in concentration of product.

feedforward regulation Stimulation of pathway by increase in concentration of precursor.

fermentation Generation of metabolites by anaerobic breakdown of glucose in some bacteria and fungi.

ferrodoxin Small soluble protein containing FeS cluster involved in electron transfer. Typically has low redox potential (strong reductant).

fosmid Cloning vector, based on bacterial F plasmid, which allows insertion of large fragments (up to 40 kb) of foreign DNA. Contain cos sites from bacteriophage lambda to allow packaging into phage. Similar to cosmid.

free fatty acids (FFA) Fatty acids derived from TAG hydrolysis in the adipose tissue and carried in the blood bound to serum albumin.

G protein Protein switch activated by GTP binding and switched off by subsequent GTP hydrolysis.

gene therapy Therapy involving introduction of recombinant DNA into organism.

glucagon Hormone used to signal fasting or starved state in animals causing up-regulation of catabolism and down-regulation of anabolism.

glucogenic Amino acids whose carbon side chain can be used to generate glucose or glycogen via gluconeogenesis.

gluconeogenesis Biosynthetic pathway by which glucose is made from pyruvate or oxaloacetate.

GLUT transporters Family of transporters for importing blood glucose into cells. Typically use facilitated diffusion, as transport is downhill.

glycerophospholipid Type of phospholipid with a glycerol backbone, esterified to a polar head group and two fatty acid chains. Also known as phosphoglyceride.

glycogen Polymer of glucose ($\alpha 1 \rightarrow 4$-linked) used for energy storage in liver and muscle of animals.

glycogen synthase Enzyme used to synthesize glycogen.

glycogenesis Biosynthetic pathway by which glycogen is synthesized from glucose.

glycogenin Polypeptide primer for glycogen synthesis.

Glossary

glycogenolysis Pathway by which glycogen is broken down to glucose 6-phosphate.

glycolysis Catabolic pathway by which one molecule of glucose is broken down to generate two molecules of pyruvate and ATP.

glycosidic bond Bond used to join sugars together in polysaccharides and to link bases to ribose sugar in DNA and RNA.

glyoxylate cycle Pathway used by plants and bacteria to make pyruvate/sugars from acetyl CoA.

glyoxysome Organelle that carries out glyoxylate cycle to enable plants to convert fatty acids to carbohydrate.

Golgi apparatus or body Organelle where most glycosylation of membrane protein and lipids takes place and proteins are sorted for different destinations.

grana Flattened thylakoid membrane vesicles arranged in stacks.

granal lamellae Membrane stacks in chloroplast in contact with other membranes of granum. Contain other enzyme complexes, e.g. photosystem II (oxygen evolution): Cyt bf complex.

guanine Purine base found in DNA and RNA.

haem Heterocyclic, conjugated ring containing Fe complexed to four internal nitrogen atoms. Found in haemoglobin and cytochromes.

H-bond Bond formed by sharing of non-bonding (lone pair) electrons from one atom with a hydrogen atom covalently attached to an electronegative atom and therefore starved of electrons. Also known as a hydrogen bond.

helicase Enzyme that disrupts the double helix of DNA using energy from ATP hydrolysis to break hydrogen bonds.

herbicide Compound that kills plants.

heterochromatin Densely packed form of chromatin, rich in non-coding DNA.

heterotrimeric Protein complex made from three different subunits.

heterotrophic Requiring a source of organic carbon for growth as unable to fix carbon (unlike autotroph).

histone code Various small chemical modifications, usually acetylation or methylation of core histone N terminal tails, which have marked effects on chromatin organization and DNA accessibility: important in regulating transcription and replication in eukaryotes. Combination of histone modifications on a nucleosome which act together to specify the fate of the associated DNA.

Holliday junction Junction formed by four strands of DNA.

homodimer Protein complex made from two identical subunits.

homologous chromosome In diploid organisms, the equivalent chromosome from either maternal or paternal source.

hormone Blood-borne messenger carrying information between tissues.

hydrodynamic volume (Stokes radius) Radius of a hard sphere which diffuses at the same rate as the molecule in question.

hydrogen bond Bond formed by sharing of non-bonding (lone pair) electrons from one atom with a hydrogen atom covalently attached to an electronegative atom and therefore starved of electrons. Also known as H bond.

hydrolase Enzyme that catalyses transfer of a group from a substrate to water as acceptor.

hydrolysis Reaction in which a complex molecule is split into smaller molecules using water as a reactant.

hydrophobic effect Interaction between hydrophobic molecules in a polar, aqueous environment to exclude water. Entropically favourable.

β-hydroxybutyrate Component of a ketone body, water-soluble form of transferring the carbon atoms from fatty acids for metabolism. See also acetone and acetoacetate.

hypoglycaemia Low blood glucose concentration.

induced fit hypothesis Model of enzyme catalysis where the active site is flexible and adjusts to fit the substrate/transition state.

inducible Able to be switched on, for example, in response to the presence of a small molecule inducer.

initial rate Rate of a reaction prior to the appearance of significant amounts of product.

initiation factors Protein factors required for initiation of translation that are not an integral part of the ribosome.

insulin Hormone used in animals to signal the fed state. Promotes uptake of glucose from blood and anabolic pathways.

integral membrane protein Protein that spans the lipid bilayer one or more times and can only be extracted by detergent.

intermediate filament Component of cytoskeleton and nuclear lamina in eukaryotic cells, used for strength.

intron Section of nucleic acid found in initial RNA transcript which is removed to form mature messenger RNA by splicing.

inverse PCR PCR which is used to amplify the DNA sequences flanking two known sequences rather than the sequences between them.

ionic bond Bond formed between two oppositely charged ions. Also known as an electrostatic bond.

irreversible inhibition Inhibition of the function of a protein by covalently attaching the inhibitor molecule.

irreversible inhibitor Becomes covalently bound to an enzyme to inhibit its activity.

irreversible reaction Reaction that can only go in one direction, with large negative ΔG.

isoelectric point pH at which a protein carries no net charge.

isoenzymes Enzymes with different amino acid sequences that catalyse the same reaction. Often have different regulatory properties.

isomerase Enzyme that catalyses an isomerization of a molecule.

isomerization Reaction in which one molecule is converted to another with the same molecular formula but a different structure.

k_{cat} Catalytic turnover constant.

ketogenic amino acid Amino acid whose carbon skeleton is broken down to acetyl CoA, but which cannot make glucose in animals.

ketone bodies Water-soluble form of the carbon atoms from fatty acids used for transport in blood, e.g. acetoacetate and β-hydroxybutyrate.

kinetics Study of rates of change.

Krebs cycle Cycle of chemical reactions by which reducing power is extracted from acetyl CoA. Also known as TCA or citric acid cycle.

lac operon Operon encoding three structural genes involved in lactose metabolism. Commonly used example of regulation of gene expression in *E. coli*.

lac repressor Protein that binds operator sequence to inhibit transcription of lac operon. DNA binding reversed by interaction with allolactose.

lagging strand Strand of DNA that is synthesized discontinuously as small Okazaki fragments, which are subsequently ligated together.

Glossary

leading strand Strand of DNA that is synthesized continuously in the same direction as replication fork movement by *DNA polymerase*.

ligase Enzyme that joins two substrates together covalently.

link reaction Pyruvate is converted into acetyl CoA to link glycolysis with the TCA cycle.

lipid Water-insoluble organic compound, usually found in cell membranes.

lipolysis Breakdown of lipid.

lipoproteins Assemblies of TAG, cholesterol, phospholipids, and proteins for blood transport of TAG. Named according to lipid content—very low density lipoprotein (VLDL) has large amounts of lipid; LDL and HDL progressively less.

lock and key hypothesis Model of enzyme catalysis where the active site is rigid and the substrate/transition state adapts to fit it.

lyase Enzyme that catalyses non-hydrolytic and non-oxidative elimination to form double bond.

lysosome Organelle in eukaryotic cells where degradative enzymes are stored. Lumen is acidic.

MAP kinases Mitogen-activated protein kinases—a family of protein kinases forming a cascade leading to activation of transcription.

metabolic channelling Series of processes in which the intermediates are held within a protein complex and not permitted to diffuse freely.

Michaelis–Menten equation Equation relating initial rate of an enzyme-catalysed reaction to substrate concentration. Predicts saturation kinetics.

microenvironment Region of a structure where the conditions (e.g. pH, water availability) differ from the bulk environment.

microfilament Filament formed by polymerization of actin which forms part of the cytoskeleton in eukaryotic cells.

microtubule Filament formed by polymerization of tubulin which forms part of the cytoskeleton in eukaryotic cells.

mitochondrion Organelle in eukaryotic cells which is major site of ATP synthesis during aerobic metabolism. Site of, for example, TCA cycle and electron transport chain.

modular Made up of many different independent units.

Monod–Wyman–Changeux model Mathematical model that predicts allosteric/cooperative behaviour of enzymes by assuming the existence of two (and only two) symmetrical forms of an assembly of several identical subunits.

mobile phase Solution flowing through a matrix in column chromatography.

monoclonal antibodies Antibody produced by a single clone of B cells in response to an antigen. All antibodies are identical and recognize the same epitope.

multicloning site Short sequence of DNA found in a plasmid or cloning vector which contains multiple sites recognised by different restriction enzymes in close proximity. Useful for insertion of foreign DNA in cloning.

mutation Permanent change in the DNA sequence (e.g. changing C to T). Has consequences if it leads to change in amino acid sequence of a protein or if it affects regulatory regions of DNA, e.g. promoters.

nitrogen fixation Reduction of atmospheric N_2 to NH_3.

nitrogenous base Base found in DNA or RNA: cytosine, guanine (G), adenine (A), thymine (T) (DNA only) and uracil (U) (RNA only).

nuclear pore complexes Protein complexes found in the nuclear membrane which allow the controlled passage of large molecules into and out of the nucleus.

nuclear matrix Proteinaceous component of nucleus involved in organization. May also contain RNA.

nucleoid Region of the cell in prokaryotes which contains the DNA, but lacking a membrane.

nucleophile Reagent that donates an electron pair to make a new bond in a chemical reaction (cf. electrophile).

nucleus Organelle of eukaryotic cells containing genomic DNA and where it is transcribed into RNA.

open reading frame Successive triplets of nucleotides that encode amino acids to be polymerized into a protein. A double-stranded molecule of DNA has six potential reading frames, three in each strand. An open reading frame contains many triplets encoding amino acids bounded by a start codon (AUG) at one end and a stop codon at the other.

operator Short sequence of DNA near promoter in bacteria that binds a protein to negatively regulate transcription.

operon Number of open reading frames close together on same strand of DNA which are transcribed, in prokaryotes, into a single polycistronic RNA for protein synthesis. Usually encode a number of proteins required in the same pathway as expression is co-regulated, e.g. lac operon.

organelle Membrane-bound structure found in eukaryotic cells with a particular function.

origin of replication Region of DNA where replication begins.

oxidation Loss of electrons by a molecule.

oxidative decarboxylation Oxidative reaction involving removal of carboxylate group to form CO_2; most commonly used to describe reaction by which pyruvate generated by glycolysis is converted into acetyl CoA, involving oxidation and decarboxylation.

oxidoreductase Enzyme that catalyses redox reactions.

PCR Polymerase chain reaction; enzymatic reaction which is used to amplify a DNA sequence between two known sequences. Involves repeated cycles of DNA melting, primer binding, and DNA synthesis (primer extension), usually using thermostable *DNA polymerases*.

pentose phosphate pathway Pathway (cycle) for interconversions of three-, four-, five-, and six-carbon sugar phosphates. May be used to generate ribose 5'-phosphate for nucleotide biosynthesis.

peptidases An enzyme that hydrolyses peptide bonds leading to protein degradation; aka protease and proteinase.

peptide bond Covalent bond between the amino group on one amino acid and the carboxyl group on another. Has partial double bond character—planar and semi-rigid. Formed on ribosome.

peptidyl transferase Enzyme that catalyses formation of peptide bond on ribosome. Made up of rRNA, not protein.

permeation limit Lower limit of molecular weight that can be resolved by size exclusion chromatography.

peripheral membrane protein Protein associated with a membrane through non-covalent interaction with phospholipid head groups or other proteins and which can be extracted by high salt.

peroxisome Vesicular organelle in eukaryotic cells that is the site of many oxidative reactions.

phosphodiester bond Covalent bond joining 3' end of one nucleotide with 5' end of another, e.g. in a chain of DNA or RNA. Synthesis catalysed by *DNA/RNA polymerases*.

phosphodiesterase Enzyme responsible for hydrolysis of cAMP to AMP.

phosphoglyceride Type of phospholipid with a glycerol backbone, esterified to a polar head group and two fatty acid chains. Also known as glycerophospholipid.

phospholipase Enzyme that cleaves bonds in a phospholipid.

Glossary

phospholipase C Enzyme that cleaves bond between diacylglycerol and phosphate in a phospholipid.

phospholipid Lipid with a polar phospho head group and hydrophobic fatty acid tails.

phosphorolysis Bond cleavage by addition of orthophosphate.

photochemistry Chemical reaction induced by photons (light).

photophosphorylation Generation of ATP from ADP coupled to the light-dependent transfer of electrons during photosynthesis.

photorespiration Rubisco-catalysed oxidation of ribulose 1,5-bisphosphate by O_2, leading to light-induced O_2 uptake by chloroplasts.

photosynthesis The process by which light energy is trapped to provide energy for living organisms.

photosystem Protein complex containing chlorophyll that can be photo-oxidized in photosynthesis to generate reducing power.

physical mapping Analysis of large regions of chromosomes that are mapped to identify loci present/absent in families with specific diseases, e.g. cystic fibrosis.

plasmid Small circular molecule of double-stranded DNA that can replicate autonomously in a cell (bacterial or lower eukaryote, e.g. yeast). Useful as cloning vectors. Often carry selectable marker, e.g. for antibiotic resistance.

polycistronic mRNA Messenger (m)RNA containing more than one open reading frame to co-ordinately direct synthesis of more than one protein in prokaryotes.

polyclonal antibodies Mixed pool of antibodies produced by different clones of B cells in an animal in response to an antigen. Will contain antibodies that recognize different epitopes of the antigen.

polymerization Joining of a large number of similar molecules together to form a polymer.

primase Enzyme that catalyses the synthesis of a short RNA molecule (primer) complementary to template DNA sequence which is then elongated by *DNA polymerase* in the replication of DNA.

primer A short oligomer (e.g. of DNA or RNA) to which an enzyme adds additional monomers. A term also used in glycogen synthesis, where the primer is the nucleation site for attachment of glucose moieties.

probe Piece of nucleic acid (usually a short single-stranded piece of DNA) that is labelled radioactively or fluorescently and which is used to locate the complementary strand of DNA or RNA by base pairing.

processivity Number of catalytic reactions of an enzyme for a single binding event. For example, *DNA polymerase* with high processivity will synthesize a long strand before detaching from its template.

product Molecule generated by a chemical reaction.

progress curve Measures the rate of increase of product [P] or decrease of substrate [S] with time.

promoter Short sequence of DNA that binds *RNA polymerase* with high affinity to generate a pre-initiation complex and direct initiation of transcription.

prosthetic group Tightly bound cofactor for an enzyme.

protease Enzyme that cleaves peptide bond to generate amino acids or short peptides from polypeptides.

protein Polymer of amino acids joined by peptide bonds.

proteinase An enzyme that hydrolyses peptide bonds leading to protein degradation; aka *protease* and *peptidase*.

protein kinase Enzyme that transfers a phosphate group from ATP to another protein. Often used in enzyme regulation.

protein phosphatase Enzyme that removes a phosphate group from proteins phosphorylated by a *protein kinase*.

proton motive force Energy stored as gradient of protons across a membrane.

pseudogene Sequence of genomic DNA that has the properties of a gene (e.g. an open reading frame and potential promoter) but which is not transcribed.

purine Two ring base found in RNA and DNA (A, G).

pyrimidine Single ring base found in RNA (U, C) and DNA (T, C).

pyrophosphatase Enzyme that catalyses the cleavage of pyrophosphate to two molecules of inorganic phosphate. Used to drive forward reactions such as polymerization of dNTPs (DNA replication) and NTPs (transcription).

pyruvate decarboxylase Enzyme that catalyses decarboxylation of pyruvate to ethanal.

rate constant (*K*) Proportionality constant relating concentration of reactants to rate of reaction.

reading frame Successive triplets of nucleotides that may encode amino acids to be polymerized into a protein. A double-stranded molecule of DNA has six potential reading frames, three in each strand.

receptor tyrosine kinase Transmembrane protein with an extracellular receptor domain (e.g. for growth factors) and an internal tyrosine kinase domain.

reduction Gain of electrons by a molecule.

regulatory enzymes Enzymes whose rate is controlled by something other than substrate or product concentration. Often allosteric.

release factor Protein that promotes termination of translation and release of the polypeptide from the ribosome.

replication fork Site of replication of double-stranded DNA.

repressible Capable of being switched off, e.g. by a small molecule repressor.

restriction site Specific sequence of bases in double-stranded DNA which is recognized by a *restriction enzyme* (*endonuclease*) that cuts the DNA backbone. Most frequently a palindrome of four or six base pairs.

retroviruses Viruses that have an RNA genome and which usually make a cDNA copy that inserts into the host (eukaryotic) genome.

reverse transcriptase (RT) *DNA polymerase* which uses an RNA template to generate a complementary DNA strand (cDNA).

reverse transcription Copying of RNA to generate a single strand of DNA by *reverse transcriptase*.

reversible inhibition Inhibition of the function of a protein by non-covalently binding the inhibitor molecule.

reversible inhibitor Inhibits enzyme activity by binding non-covalently so it can be removed.

ribonucleic acid Single-stranded polymer of ribonucleotides linked by phosphodiester bonds. Also known as RNA.

Ribonucleoside Nitrogenous base linked to the C1 carbon of 2′ ribose by a β-N-glycosidic bond, i.e. base + sugar.

Ribonculeotide Ribonucleoside with the 5′ carbon of ribose joined by an ester linkage to a phosphate group (NMP) with or without additional phosphates (NDP) (NTP), i.e. base + sugar phosphate(s).

Glossary

ribosome Protein/RNA complex that catalyses synthesis of protein directed by mRNA. Consists of a small and large subunit.

ribozyme Enzyme made of RNA.

RNA Single-stranded polymer of ribonucleotides linked by phosphodiester bonds. Also known as ribonucleic acid.

RNA polymerase Enzyme that catalyses formation of phosphodiester bonds between NTPs to generate RNA. Always synthesizes RNA in 5′ to 3′ direction. Requires a DNA template as well as NTPs.

Rubisco *Ribulose 1,5-bisphosphate carboxylase/oxygenase*—enzyme responsible for primary CO_2 fixation in plants.

saturated (enzyme) Situation where all active sites are occupied, such that increasing substrate concentration does not affect rate.

saturated (fats) All carbons are linked by single covalent bonds (two hydrogens per C within chain, third H linked to the terminal C).

second messenger Small molecule/ion which acts as an intracellular intermediate to carry information from a hormone receptor to a target enzyme.

selectable marker Gene that codes for a protein which allows the cell expressing the protein to be distinguished from other cells, e.g. promotes cell survival in certain conditions such as drug resistance.

semiconservative replication Mode of replication in which phosphodiester backbone of each strand of DNA remains intact and acts as template for a new strand.

Shine–Dalgarno sequence Short sequence of RNA found upstream of an open reading frame in prokaryotes which acts as a ribosome binding site by base pairing with 16S rRNA.

shuttle vector Plasmid or cloning vector that can function in cells from more than one host organism.

sigmoidal S-shaped. Cooperativity leads to sigmoid binding curves.

sister chromatids Two identical copies of a single chromosome connected at a centromere

sphingolipid Lipid with backbone of sphingosine with fatty acyl chains attached.

spliceosome Complex multiprotein–RNA assembly with multiple snRNPs in nucleus of eukaryotes, which carries out splicing (i.e. removal of introns from primary transcripts).

splicing Removal of introns to generate mature messenger RNA.

starch Polymer of glucose serving as a compact store of energy in plants. Exists as mix of two forms: amylose (α-1,4 linkages) and amylopectin (like amylose but with a branch with an α-1,6 linkage approximately every 30 residues).

stationary phase Resin or gel of, for example, agarose or sepharose modified to associate with particular molecules in a mixed sample in column chromatography.

steady state assumption Approximation that during a complex process/pathway, the levels of the intermediates remain constant. Can be used to derive the Michaelis–Menten equation.

stereoisomer Molecules with the same molecular formula and functional groups but which differ only in the 3-D arrangement of the atoms in space.

stringent Hybridization or washing conditions used to identify molecules most tightly binding probes: often low salt and/or high temperature.

stroma Aqueous compartment inside a chloroplast.

stroma lamellae Membrane stacks in chloroplast in contact with stroma and containing various enzyme complexes, e.g. photosystem I: *ATP synthase*.

substrate Reactant in enzyme-catalysed reaction. Converted to product.

substrate specificity Ability of an enzyme to select one substrate over another.

supersecondary structure Arrangement of secondary structural elements found commonly amongst proteins (e.g. Ig fold, $\alpha\beta$ barrel). Often forms a framework on which the active site is placed.

symbiotic Close association between different species for mutual benefit.

synthase Enzyme that catalyses the addition of one substrate to a double bond of another.

synthetase Enzyme that catalyses the ligation of two substrates and requires energy from NTP hydrolysis.

TCA cycle Cycle of chemical reactions by which reducing power is extracted from acetyl CoA. Also known as Krebs or citric acid cycle.

telomere DNA loop structure comprising tandem repeats, bound by a multi-protein complex, to stabilize the ends of linear chromosomes in eukaryotic cells.

thioredoxin Small protein containing two cysteine side chains that can exist in reduced (2SH) or oxidized (S–S) forms.

thylakoid membranes Internal membranes of chloroplasts. Contain photosystems and antennae chlorophyll.

thymine Pyrimidine base found only in DNA (uracil in RNA).

transamination Transfer of an amine group from an amino acid to another for subsequent safe excretion (via conversion into urea in animals) or amino acid synthesis.

transcription Copying of one strand of DNA to generate a single strand of RNA, catalysed by *RNA polymerase*.

transcription bubble Short section of DNA unwound to allow transcription.

transcription factor Protein that regulates ability of *RNA polymerase* to transcribe a particular subset of genes.

transfection Experimentally induced uptake of DNA into eukaryotic cells.

transferase Enzyme that catalyses transfer of a group from one substrate to another.

transformation Uptake of foreign DNA into *E. coli*.

transition state Least stable species on the path between substrate and product. Stabilized by enzyme.

translation RNA-directed polymerization of amino acids by peptide bond formation to generate a protein. Takes place on the ribosome.

translocase Enzyme that moves a molecule.

transposon A piece of DNA that can move to a different position in the genome.

treadmilling The rate of addition of monomers to one end of a filament is equal to the rate of depolymerization at the other, so the overall length does not change.

triacylglycerol (TAG) Glycerol with one fatty acid chain esterified to each of its three OH groups. Also known as triglyceride.

triglyceride Glycerol with one fatty acid chain esterified to each of its three OH groups. Also known as triacylglycerol.

turnover number (k_{cat}) Maximum number of molecules of substrate that can be converted into product by one molecule of enzyme per unit time.

ubiquitin A small (76 amino acid) highly conserved eukaryotic protein that can be ligated to target proteins; multiple ubiquitin molecules ligated in tandem target a protein for degradation by the proteasome; highly conserved from yeast to humans.

uncoupling proteins Proteins that span the inner mitochondrial membrane, allowing passage of H^+ down the concentration gradient without generation of ATP; play a role in heat generation.

unsaturated Containing one or more double bond.

Glossary

uracil (U) Pyrimidine base found only in RNA (thymine in DNA).

urea Used to safely excrete ammonia in animals.

urea cycle Cycle of reactions used to convert ammonia to urea for excretion in animals.

van der Waals force/attraction Weak attractive force between molecules. Transient dipoles formed by electron movement.

very low density lipoprotein (vLDL) 30–80 nm particle produced in liver from cholesterol, triglycerides and apolipoproteins; subsequently converted in bloodstream to LDL.

vitamin Complex organic molecule that needs to be provided in the diet of humans. Often used as cofactor in enzyme reactions.

void volume Volume of mobile phase outside porous beads in a column in column chromatography.

yeast artificial chromosomes Linear artificial chromosomes that can replicate and segregate correctly during the cell division process in yeast, having yeast centromeres and telomeres as well as replication origins. Used to clone large fragments of DNA.

zwitterion Molecule containing both a positive charge and a negative charge, e.g. an amino acid in which the amine group is positively charged and the carboxylic acid group is negatively charged at neutral pH.

Index

Index

Index

Index

Index